Emerging Trends in Immunomodulatory Nanomaterials Toward Cancer Therapy

Synthesis Lectures on Biomedical Engineering

Editor
John D. Enderle, *University of Connecticut*

Lectures in Biomedical Engineering will be comprised of 75- to 150-page publications on advanced and state-of-the-art topics that span the field of biomedical engineering, from the atom and molecule to large diagnostic equipment. Each lecture covers, for that topic, the fundamental principles in a unified manner, develops underlying concepts needed for sequential material, and progresses to more advanced topics. Computer software and multimedia, when appropriate and available, are included for simulation, computation, visualization and design. The authors selected to write the lectures are leading experts on the subject who have extensive background in theory, application and design.

The series is designed to meet the demands of the 21st century technology and the rapid advancements in the all-encompassing field of biomedical engineering that includes biochemical processes, biomaterials, biomechanics, bioinstrumentation, physiological modeling, biosignal processing, bioinformatics, biocomplexity, medical and molecular imaging, rehabilitation engineering, biomimetic nano-electrokinetics, biosensors, biotechnology, clinical engineering, biomedical devices, drug discovery and delivery systems, tissue engineering, proteomics, functional genomics, and molecular and cellular engineering.

Emerging Trends in Immunomodulatory Nanomaterials Toward Cancer Therapy
Anubhab Mukherjee, Vijay Sagar Madansetty, and Sudip Mukherjee
2020

Nanotechnology for Bioengineers
Wujie Zhang
2020

Fast Quantitative Magnetic Resonance Imaging
Guido Buonincontri, Joshua Kaggie, and Martin Graves
20208

3D Electro-Rotation of Single Cells
Liang Huang and Wenhui Wang2017

Emerging Trends in Immunomodulatory Nanomaterials Toward Cancer Therapy
Anubhab Mukherjee, Vijay Sagar Madamsetty, and Sudip Mukherjee

ISBN: 978-3-031-00541-1 print
ISBN: 978-3-031-01669-1 ebook
ISBN: 978-3-031-00048-5 hardcover

DOI 10.1007/978-3-031-01669-1

A Publication in the Springer Nature series
SYNTHESIS LECTURES ON ADVANCES IN AUTOMOTIVE TECHNOLOGY
Lecture #61
Series Editor: John D. Ederle, University of Connecticut

Series ISSN 1930-0328 Print 1930-0336 Electronic

Emerging Trends in Immunomodulatory Nanomaterials Toward Cancer Therapy

Anubhab Mukherjee
Esperer Onco Nutrition Pvt. Ltd.

Vijay Sagar Madamsetty
Department of Biochemistry and Molecular Biology, Mayo Clinic College of Medicine and Science

Sudip Mukherjee
Department of Bioengineering, Rice University

SYNTHESIS LECTURES ON BIOMEDICAL ENGINEERING #61

ABSTRACT

Recently, immunomodulatory nanomaterials have gained immense attention due to their involvement in the modulation of the body's immune response to cancer therapy. This book highlights various immunomodulatory nanomaterials (including organic, polymer, inorganic, liposomes, viral, and protein nanoparticles) and their role in cancer therapy. Additionally, the mechanism of immunomodulation is reviewed in detail. Finally, the challenges of these therapies and their future outlook are discussed. We believe this book will be helpful to a broad community including students, researchers, educators, and industrialists.

KEYWORDS

immunomodulatory nanomaterials, cancer therapy, clinical trials, immunotherapy, nanomedicine

Contents

CHAPTER 1

Introduction

Anubhab Mukherjee, Vijay Sagar Madamsetty, Sudip Mukherjee

1.1 CANCER: A GLOBAL DISEASE, STATISTICS

Global biomedical research would indubitably agree that cancer is one of the biggest threats to human survival causing severe human demise on the planet. Second to cardiovascular diseases, cancer causes enormous health, social, and economic burdens. As estimated by the World Health Organization (WHO) and American Cancer Society, approximately 17 million cancer incidences occurred and 9.6 million people died worldwide due to cancer throughout the world in 2018 alone. In 2019, an estimated 1,762,450 new cancer cases were reported and approximately 606,880 cancer deaths happened in the U.S. In 2020, 1,806,590 new cancer cases and 606,520 cancer deaths occurred in the U.S. In 2021, 1,898,160 new cancer cases and 608,570 cancer related demises are projected to take place in the U.S. as well. This hints at a traumatic fact that every day there will be ~5,200 cancer cases diagnosed in the U.S., let alone the world. By 2040, the global incidence of cancer will reach 27.5 million newly diagnosed each year, which is a catastrophic state of affair in itself [1, 2]. The rate of human demise by cancer was heightened until 1991, then slopped down continuously through 2017, effectuating an overall decline of 29% that translates into an estimated 2.9 million fewer cancer deaths than would have occurred if peak rates had persisted. Four leading cancers, namely—lung, colorectal, breast, and prostate—have encountered this prolonged decline in death rate. Even so, over the last decade, diminutions decelerated for female breast and colorectal cancers, and almost stopped for prostate cancer. Contrastingly, the decline in death rate was improved for lung cancer, from 3%–5% annually during 2008–2017 in men and from 2%–4% in women, instigating the largest ever single-year drop in overall cancer mortality of 2.2% from 2016–2017. Yet, 228,820 new lung cancer cases and 135,720 deaths were predicted in the U.S. in 2020, and it still remains the leading cause of global cancer mortalities (18.4% of total cancer fatalities). Conventional cancer therapies have several limitations and associated side effects. As exhibited by NCHS, WHO, and NCI, early detection and diagnosis coupled with superior treatment modalities can alleviate the fatality of cancer [3, 4]. We shall discuss this in the next section.

1.2   CONVENTIONAL TREATMENT STRATEGIES AND CHALLENGES

For the last several decades, chemotherapy has remained the preeminent modality for the treatment of cancer [5, 6]. The first kind of chemotherapeutics tested in humans were chlorambucil and cyclophosphamide, which induced cellular cytotoxicity by DNA alkylation. The observation that folic acid instigates cancer cell growth led to the synthesis of antifolates, such as methotrexate, as antitumor agents. Subsequently, nucleoside analogues, such as 5-fluoruracil and cytosine arabinoside (ara-C), which essentially interfere with DNA synthesis, were synthesized. This was followed by the development of DNA inter-chelating agents (viz., cisplatin, actinomycin D, etc.), anthracyclines (viz., Doxorubicin, etc.), and tubulin targeting molecules (viz., Taxanes, Vinca alkaloids, etc.). The vivid triumph for the extinction of cancer cells was inexorably equaled by the monumental failure of tumor selectivity, which ended up with killing their normal counterparts, eventually leading to huge untoward systemic toxicities. Nonetheless, it was soon ostensibly recognized that therapeutics with non-overlapping toxicity profiles and different mechanisms of action could be combined at different dose regimens to yield synergistic effects with ameliorated antitumor activity. The combinatorial chemotherapy thus emerged as an ensuing modality for the treatment of most cancers [7]. Ardent votaries in the clinic-defined maximum tolerated dose (MTD), minimum effective dose (MED), and later metronomic dose (MET), and worked with a plethora of individual drugs and their combinations in order to reach an improved therapeutic index.

In the last two decades of the present century, surpassed understanding of cancer biology has spawned a new era of targeted cancer treatment by exploring a few distinguishing properties of tumor cells in contrast to normal cells [8]. These hallmark capabilities enable tumor cells to survive, multiply, and metastasize by altering several cellular mechanics, which include: obtaining an autonomy in growth signals, insensitivity to anti-growth signals, evasion of apoptosis, infinite replicative potential, sustained angiogenesis, tissue invasion and metastasis, reprogramming cellular metabolism, and evasion of immunological destruction. Therapeutic targeting of the "hallmarks of cancer" provides safer and more efficacious alternatives to traditional chemotherapy. Stated differently, targeted (tissue, cellular, molecular, etc.) therapy promises selective cytotoxicity to tumor cells, ensuring overall lower toxicity to the host, yielding a higher therapeutic index.

Importantly, cancer cells differ from normal cells due to genomic instability and thus mutations in oncogenes and/or tumor suppressor genes [9]. When the integrity of the genome is compromised, cells are more likely to develop additional genetic faults, some of which may give rise to tumor-specific antigens (found only on the surface of tumor cells) or tumor-associated antigens (overexpressed on tumor cells, but also present on normal cells) [10]. Current scientific endeavor has led to the discovery that several human cancers express unique tumor-specific or tumor-associated cell surface antigens, which are of great value as targets for monoclonal antibody (mAb)-based therapy [11]. Also, an entirely new class of tripartite anti-cancer drugs, namely, antibody-drug conjugates (ADCs), have come into play and a few of them are already commercialized in the recent past [12]. These are comprised of a tumor-specific mAb conjugated to a potent cytotoxin via a "stable" linker.

Like normal tissues, tumors nourish themselves by providing food and oxygen as well as by developing a capacity to remove metabolic wastes and CO_2. Unarguably, tumor-associated neovascularization, obtained by angiogenesis, meet these demands. Angiogenesis, the sprouting of new capillaries from existing ones, remains almost always turned on during the process of tumorigenesis for sustenance of neoplastic expansions [13–15]. The conceptualization of decoupling tumors from surrounding blood vessels has spawned discovery and clinical approval of several anti-angiogenic drugs, namely monoclonal antibody inhibitors (Bevacizumab, IMC-1121B, 2C3), receptor tyrosine kinase inhibitors (sorafenib, sunitinib, pazopanib), soluble receptor chimeric protein (VEGF-Trap), inhibitors of endothelial cell proliferation (thalidomide, angiostatin), inhibitors of integrin's proangiogenic activity (Cilengitide, medi-522), matrix metalloproteinase inhibitors (Neovastat, Prinomastat, Marimastat), vascular targeting drug (combretastatin), etc. [16–19]. Still, attempts of anti-angiogenic cancer therapy alone or in combination with conventional chemotherapy or radiotherapy failed to combat tissue invasion and metastases. The progress of antiangiogenic cancer therapy was retarded due to drug resistance, upregulation of various proangiogenic signals, hypoxia resistance, delayed response to radiotherapy, toxicity issues, etc. [20–23].

Dietary supplements have also demonstrated their potential to be used as adjuvant therapies and have become a part of the zeitgeist. This can be illustrated with an example. The limitless replicative potential of cells that represents the essence of neoplastic disease involves not only deregulated control of cell proliferation but also corresponding adjustments of energy metabolism to sustain cell growth and division. Under aerobic conditions, normal cells break down glucose, first to pyruvate via glycolysis in the cytosol and thereafter to carbon dioxide in the mitochondria. Under anaerobic conditions, glycolysis is favored and relatively little pyruvate is dispatched to the oxygen-consuming mitochondria. Nobel laureate Otto Warburg first observed a metabolic switch, an anomalous characteristic of energy metabolism in cancer cell: even in the presence of oxygen, cancer cells reprogram their glucose metabolism and thus their energy (ATP) production, by limiting their energy metabolism largely to glycolysis instead of mitochondrial oxidative phosphorylation, producing a high level of lactate and leading to a state that has been termed "aerobic glycolysis" [24, 25]. In addition, owing to mitochondrial dysfunction and down-regulation of enzymes necessary for ketone utilization, some cancers fail to metabolize ketone bodies. Targeting the Warburg effect, the Ketogenic diet (KD, a high-fat/low-carbohydrate/adequate-protein diet) has recently been proposed as an adjuvant therapy in cancer treatment. The rationale behind providing KD, almost equaled to fasting, is to reduce the level of circulating glucose, induce ketosis so that cancer cells are starved of energy, and to continue to nourish normal cells to adapt their metabolism to use ketone bodies and survive. Moreover, a diminished level of blood glucose also curtails the levels of insulin and insulin-like growth factor, which are crucial drivers of cancer cell proliferation [26, 27]. As we know, glucose instigates pancreatic β cells to secret insulin, which allows glucose to enter cells and provide energy. An enhanced carbohydrate and glucose intake, thus, makes the pancreas release more insulin, which promotes the interaction of growth hormone and growth hormone receptors to produce insulin-like growth factor 1 (IGF-1) in the liver, in turn, promoting cell growth and proliferation, which can be pernicious to cancer patients. Positron emission tomography (PET) study reveals an overexpression of glucose transporters 1 and 3 (Glut-1, Glut-3) which corresponds to the degree of glucose uptake in aggressive tumors [28, 29]. Furthermore, translocation of upregulated hexokinase, a rate-limiting enzyme of glycolysis, from the cytosol to the outer mitochondrial membrane, sever caspase-dependent cytochrome release evading the apoptotic pathways and rendering cancer cells impervious to chemotherapy. For KD (fat to carbohydrate ratio ~4:1), with nominal availability of glucose, the body inclines toward an alternative form of energy for cells. The liver responds back producing ketone bodies (such as β-hydroxybutyrate) from fatty acids and provides energy to peripheral tissues. Eventually, glucose-dependent tumor cells continue to starve under KD impeding insulin/IGF and downstream signal transduction pathways, such as phosphoinositide 3-kinase (PI3K)/protein kinase B (Akt)/mammalian target of rapamycin (mTOR). Ketogenic diets also amplify adenosine monophosphate-activated protein kinase (AMPK), which inhibits aerobic glycolysis and suppresses tumor proliferation, invasion, and migration.

Conventional cancer therapies, including surgery, chemotherapy, and radiation therapy, are long being regularly used in the treatment of the patients with cancer and have already faced severe side-effects. One of the main problems of conventional cancer therapy is the low specificity of chemotherapeutic drugs for cancer cells. In fact, most drugs act both on healthy and diseased tissues, ending up with killing the normal cells and enhancing systemic toxicity. There is no denying the fact that despite crucial advances, the presently adopted approaches to cancer treatment are still reductionist. Targeting single molecular abnormalities in genome or cancer pathways has achieved good clinical responses that have modestly affected survival in some cancers. However, targeting a single hallmark or pathway with a single drug ("magic bullet") will not likely lead to cancer cure. A combinatorial platform where targeted therapy and immunotherapy can be coupled in a well-orchestrated fashion can address multiple issues. In the following section, we shall focus on the advent of nanomedicine as a compelling biomedical tool for circumventing challenges of the conventional therapeutic modalities [30].

1.3   ALTERNATIVE TREATMENT: NANOMEDICINE

Nanomaterials (NMs) are engineered chemical substances or materials with a particle size of 1–100 nm in diameter [31]. These NMs are extensively explored and engaged for marketable purposes in different fields, where many of the sophisticated NMs are at the important rim of promising sectors of biotechnology and biomedicine. They display inimitable physicochemical attributes, which are recognized to their diameter in nanometer, lofty surface area, surface charge, composition, morphology, and surface composition [32]. Because of their remarkably specific physicochemical properties, NMs are significantly different from their own bulk materials, allowing them to perform well with improved functionality, sensitivity, competence, and selectivity toward developing biomedicines [33]. A variety of NMs, including metal nanoparticles, liposomes, quantum dots, polymeric micelles, dendrimers, and carbon-based nanoparticles, has been used to get desired biomedical applications [34]. Two key mechanisms for drug-loaded NMs to get into the diseased sites are passive targeting and active targeting. A passive targeting mechanism happens via enhanced permeability and retention (EPR) [35]. In an active targeting mechanism, the surface functionalized NMs with various biomarkers, which specifically bind with receptors over-expressed at the pathological tissue [36,37].

In recent years, nanotechnology (NT) and associated disciplines have gained rapid escalation in biomedical implementations such as nanomedicine, solar energy, and electronics [38,39]. Biomedicine embraces the design and synthesis of NMs, along with other nanoparticles (NPs) and nano-devices [40]. NMs play a very crucial role in biomedicine including diagnosis, testing, tracking, and treating various diseases, in addition to their camaraderie with biological entities [41,42]. Once properly formulated, they show their natural aptitude to traverse with blood flow of the living entities via various routes based on their attributes and eventually get access to all the organs [43,44]. Due to their intrinsic biocompatible interactions, the NMs might have tunable physicochemical attributes, which is furthermore linked with their lesser immunogenicity and non-toxicity [45,46]. There are numerous advantages of using NMs for various biological applications: (i) it increases the concentration of drug in the pathological tissues and control the slow release of the drug; (ii) it solves issues connected to the poor solubility and bioavailability of the drug; and (iii) enhanced biodegradability and biocompatibility (iv) drugs/genes/imaging agents can be easily loaded due to their tunable surface functionalities [47,48]. Imaging agents could endow *in vivo* drug tracking ability to determine drug delivery efficacy during treatment. In recent years, various nanoparticles such as liposomes, polymers, metal nanoparticles, and inorganic nanoparticles have been developed which are able to selectively target tumor cells without causing any destruction to healthy cells or organs [49]. This scientific background will help us understand various roles that immune system plays in cancer, the biology of immune response in cancer therapy. So, we proceed to the next chapter.

1.4   REFERENCES

1.   Bray, F., Ferlay, J., Soerjomataram, I., Siegel, R. L., Torre, L. A., and Jemal, A. (2018). Global cancer statistics 2018: GLOBOCAN estimates of incidence and mortality worldwide for 36 cancers in 185 countries, *CA Cancer J. Clin.* 68 394-424 DOI: 10.3322/caac.21492. 1

2.   Siegel, R. L., Miller, K. D., and Jemal, A. (2020). Cancer statistics, *CA Cancer J. Clin.* 70:7-30 DOI: 10.3322/caac.21590. 1

3.   Siegel, R. L., Miller, K. D., and Jemal, A. (2019). Cancer statistics, *CA Cancer J. Clin.* 69:7–34.DOI: 10.3322/caac.21551. 1

4.   Wild, C. P. (2019). The global cancer burden: Necessity is the mother of prevention, *Nat. Rev. Cancer,* 19:123–124. DOI: 10.1038/s41568-019-0110-3. 1

5.   DeVita, Jr., V. T. and Chu, E. (2008). A history of cancer chemotherap, *Cancer Res.* 68:8643–8653. DOI: 10.1158/0008-5472.CAN-07-6611. 2

6.   DeVita, Jr., V. T. (1978). The evolution of therapeutic research in cancer, *N. Engl. J. Med.* 298 907–910. DOI: 10.1056/NEJM197804202981610. 2

7.   Hanna, N. H. and Einhorn, L. H. (2014). Testicular cancer-discoveries and updates, *N. Engl. J. Med.* 371 2005–2016. DOI: 10.1056/NEJMra1407550. 2

8.   Hanahan, D. and Weinberg, R. A. (2011). Hallmarks of cancer: the next generation, *Cell* 144, 646–674. DOI: 10.1016/j.cell.2011.02.013. 2

9.   Luo, J., Solimini, N. L., and Elledge, S. J. (2009). Principles of cancer therapy: oncogene and non-oncogene addiction, *Cell* 136:823–837. DOI: 10.1016/j.cell.2009.02.024. 2

10.   Sensi, M. and Anichini, A. (2006). Unique tumor antigens: evidence for immune control of genome integrity and immunogenic targets for T cell-mediated patient-specific immunotherapy, *Clin. Cancer Res.* 12:5023–5032. DOI: 10.1158/1078-0432.CCR-05-2682. 2

11.   Vigneron, N., Stroobant, V., Van den Eynde, B. J., and van der Bruggen, P. (2013). Database of T cell-defined human tumor antigens: the 2013 update, *Cancer Immun.* 13:15. 3

12.   Mukherjee, A., Waters, A. K., Babic, I., Nurmemmedov, E., Glassy, M. C., Kesari, S., and Yenugonda, V. M. (2019b). Antibody drug conjugates: Progress, pitfalls, and promises *Hum. Antibodies,* 27:53–62 DOI: 10.3233/HAB-180348. 3

13.   Hanahan, D. and Folkman, J. (1996). Patterns and emerging mechanisms of the angiogenic switch during tumorigenesis, *Cell,* 86:353–364. DOI: 10.1016/S0092-8674(00)80108-7. 3

14.   Dvorak, H. F. (2005). Angiogenesis: Update 2005. *J. Thromb. Haemost.* 3:1835–1842. DOI: 10.1111/j.1538-7836.2005.01361.x. 3

15.   Folkman, J. (1995). Angiogenesis in cancer, vascular, rheumatoid and other disease, *Nat. Med.* 1:27–31. DOI: 10.1038/nm0195-27. 3

16.   Chung, A. S, Lee, J., and Ferrara, N. (2010). Targeting the tumor vasculature: Insights from physiological angiogenesis, *Nat. Rev. Cancer* 10:505–514. DOI: 10.1038/nrc2868. 3

17.   Prager, G.W., Poettler, M., Unseld, M., and Zielinski, C. C. (2012). Angiogenesis in cancer: Anti-VEGF escape mechanisms, *Transl. Lung Cancer Res.* 1:14–25. DOI: 10.3978/j.issn.2218-6751.2011.11.02. 3

18.   Clarke, S. J. and Sharma, R. (2006). Experimental & clinical pharmacology: Angiogenesis inhibitors in cancer mechanisms of action, *Aust. Prescr.* 29:9–12. DOI: 10.18773/austprescr.2006.007. 3

19.   Ricart, A. D., Ashton, E. A., Cooney, M. M., Sarantopoulos, J., Brell, J.M., Feldman, M.A., Ruby, K.E., Matsuda, K., Munsey, M. S., Medina, G., Zambito, A., Tolcher, A. W., and Remick, S. C. (2011). A phase I study of MN-029 (denibulin), a novel vascular disrupting agent, in patients with advanced solid tumors, *Cancer Chemother. Pharmacol.* 68 959–970. DOI: 10.1007/s00280-011-1565-4. 3

20.   Ebos, J. M. L., Lee, C. R., and Kerbel, R. S. (2009). Tumor and host-mediated pathways of resistance and disease progression in response to antiangiogenic therapy, *Clin. Cancer Res.* 15:5020–5025. DOI: 10.1158/1078-0432.CCR-09-0095. 3

21.   Abdalla, A. M. E., Xiao, L., Ullah, M. W., Yu, M., Ouyang, C., and Yang, G. (2018). Current challenges of cancer antiangiogenic therapy and the promise of nanotherapeutics, *Theranostics* 8:533–548. DOI: 10.7150/thno.21674. 3

22.   Mukherjee, S. (2018). Recent progress toward antiangiogenesis application of nanomedicine in cancer therapy, *Future Sci. OA* 4. DOI: 10.4155/fsoa-2018-0051. 3

23.   Mukherjee, A., Madamsetty, V. S., Paul, M. K., and Mukherjee, S. (2020). Recent Advancements of nanomedicine toward antiangiogenic therapy in cancer, *Int. J. Mol. Sci.* 21:455. DOI: 10.3390/ijms21020455. 3

24.   Vander Heiden, M. G., Cantley, L. C., and Thompson, C. B. (2009). Understanding the Warburg effect: the metabolic requirements of cell proliferation, *Science* 324:1029–1033. DOI: 10.1126/science.1160809. 3

25.   Kennedy, K. M. and Dewhirst, M. W. (2010). Tumor metabolism of lactate: the influence and therapeutic potential for MCT and CD147 regulation, *Future Oncol.* 6:127–148. DOI: 10.2217/fon.09.145. 3

26.   Weber, D. D., Aminzadeh-Gohari, S., Tulipan, J., Catalano, L., Feichtinger, R. G., and Kofler, B. (2020). Ketogenic diet in the treatment of cancer—Where do we stand?, *Mol. Metab.* 33:102-121 DOI: 10.1016/j.molmet.2019.06.026. 4

27.   Woolf, E. C., Syed, N., and Scheck, A. C. (2016). Tumor metabolism, the ketogenic diet and b-hydroxybutyrate: Novel approaches to adjuvant brain tumor therapy. *Front. Mol. Neurosci.* 9:122. DOI: 10.3389/fnmol.2016.00122. 4

28.   Voss, M., Lorenz, N. I., Luge, A.-L., Steinbach, J. P., Rieger, J., and Ronellenfitsch, M. W. (2018). Rescue of 2-deoxyglucose side effects by ketogenic diet, *Int. J. Mol. Sci.* 19:2462. DOI: 10.3390/ijms19082462. 4

29.   de Groot, S., Pijl, H., van der Hoeven, J. J. M., and Kroep, J. R. (2019). Effects of short-term fasting on cancer treatment, *J. Exp. Clin. Cancer Res.* 38(1):209 DOI: 10.1186/s13046-019-1189-9. 4

30.   Zugazagoitia, J., Guedes, C., Ponce, S., Ferrer, I., Molina-Pinelo, S., and Paz-Ares, L. (2016). Current challenges in cancer treatment, *Clin. Ther.* 38(7):1551-66 DOI: 10.1016/j.clinthera.2016.03.026. 4

31.   Pandya, T., Patel, K. K., Pathak, R., and Shah, S. (2019). Liposomal formulations in cancer therapy: Passive versus active targeting, *Asian J. Pharm. Res. Dev.* 7:35–38. DOI: 10.22270/ajprd.v7i2.489. 4

32.   Boverhof, D. R., Bramante, C. M., Butala, J. H., Clancy, S. F., Lafranconi, M., West, J., and Gordon, S. F. (2015). Comparative assessment of nanomaterial definitions and safety evaluation considerations, *Regul. Toxicol. Pharmacol.* 73:137–150. DOI: 10.1016/j.yrtph.2015.06.001. 5

33.   Jain, K. K. (2008). Nanomedicine: Application of nanobiotechnology in medical practice. *Med. Princ. Pract.* 17:89–101. DOI: 10.1159/000112961. 5

34.   Kawasaki, E. S. and Player, A. (200). Nanotechnology, nanomedicine, and the development of new, effective therapies for cancer. *Nanomed. Nanotechnol., Biol. Med.* 1:101–109. DOI: 10.1016/j.nano.2005.03.002. 5

35.   Golombek, S. K., May, J. N., Theek, B., Appold, L., Drude, N., Kiessling, F., and Lammers, T. (2018). Tumor targeting via EPR: Strategies to enhance patient responses, *Adv. Drug Deliv. Rev.* 130:17–38. DOI: 10.1016/j.addr.2018.07.007. 5

36.   Mukherjee, S., Madamsetty, V. S., Bhattacharya, D., Chowdhury, S. R., Paul, M. K., and Mukherjee, A. (2020). Recent advancements of nanomedicine in neurodegenerative disorders theranostics, *Adv. Funct. Mater.* 30:2003054. DOI: 10.1002/adfm.202003054. 5

37. Sanna, V. and Sechi, M. (2020). Therapeutic potential of targeted nanoparticles and per-spective on nanotherapies, *ACS Med. Chem. Lett.* 11:1069–1073. DOI: 10.1021/acsmed-chemlett.0c00075. 5

38. Mohapatra, S. S., Frisina, R. D., Mohapatra, S., Sneed, K. B., Markoutsa, E., Wang, T., Dutta, R., Damnjanovic, R., Phan, M.-H., Denmark, D. J., Biswal, M. R., McGill, A. R., Green, R., Howell, M., Ghosh, P., Gonzalez, A., Ahmed, N. T., Borresen, B., Farmer, M., Gaeta, M., Sharma, K., Bouchard, C., Gamboni, D., Martin, J., Tolve, B., Singh, M., Judy, J. W., Li, C., Santra, S., Daunert, S., Zeynaloo, E., Gelfand, R. M., Lenhert, S., McLam-ore, E. S., Xiang, D., Morgan, V., Friedersdorf, L. E., Lal, R., Webster, T. J., Hoogerheide, D. P., Nguyen, T. D., D'Souza, M. J., Çulha, M., Kondiah, P. P. D., and Martin, D. K., (2020). Advances in translational nanotechnology: Challenges and opportunities, *Appl. Sci.* 10:4881. DOI: 10.3390/app10144881. 5

39. Zhang, F. (2017). Grand challenges for nanoscience and nanotechnology in energy and health. *Front. Chem.* 5 80. DOI: 10.3389/fchem.2017.00080. 5

40. Muthu, M. S., Leong, D. T., Mei, L., and Feng, S.S. (2014). Nanotheranostics - appli-cation and further development of nanomedicine strategies for advanced theranostics, *Theranostics* 4:660–677. DOI: 10.7150/thno.8698. 5

41. Zahin, N., Anwar, R., Tewari, D., Kabir, T., Sajid, A., Mathew, B., Uddin, S., Aleya, L., and Abdel-Daim, M. M. (2020). Nanoparticles and its biomedical applications in health and diseases: special focus on drug delivery, *Environ. Sci. Pollut. Res.* 27:19151–19168. DOI: 10.1007/s11356-019-05211-0. 5

42. Bhushan, I., Singh, V. K., and Tripathi, D. K. (2020). *Nanomaterials and Environmental Biotechnology*. Springer International Publishing. DOI: 10.1007/978-3-030-34544-0. 5

43. Sims, C. M., Hanna, S. K., Heller, D. A.,Horoszko, C. P., Johnson, M. E., Montoro Bus-tos, A. R., Reipa, V., Riley, K. R., and Nelsom, B. C. (2017). Redox-active nanomaterials for nanomedicine applications, *Nanoscale* 9:15226–15251. DOI: 10.1039/C7NR05429G. 5

44. Fadeel, B., Bussy, C., Merino, S., Vázquez, E., Flahaut, E., Mouchet, F., Evariste, L., Gauthier, L., Koivisto, A. J., Vogel, U., Martin, C., Delogu, L. G., Buerki-Thurnherr, T., Wick, W., Beloin-Saint-Pierre, D., Hischier, R., Pelin, M., Carniel, F. C., Tretiach, M., Cesca, F., Benfenati, F., Scaini, D., Ballerini, L., Kostarelos, K., Prato, M., and Bianco, A., (2018). Safety assessment of graphene-based materials: focus on human health and the environment, *ACS Nano.* 12:10582–10620. DOI: 10.1021/acsnano.8b04758. 5

45.   Navya, P. N. and Daima, H. K. (2016). Rational engineering of physicochemical properties of nanomaterials for biomedical applications with nanotoxicological perspectives, *Nano Converg.* 3:1. DOI: 10.1186/s40580-016-0064-z. 5

46.   Mu, Q., and Yan, B. (2019). Editorial: Nanoparticles in cancer therapy-novel concepts, mechanisms and applications, *Front. Pharmacol.* 91552. DOI: 10.3389/fphar.2018.01552. 5

47.   Yadollahi, R., Vasilev, K., and Simovic, S. (2015). Nanosuspension technologies for delivery of poorly soluble drugs, *J. Nanomater.* 2015. DOI: 10.1155/2015/216375. 5

48.   Sharma, M., Sharma, R., and Jain, D. K. (2016). Nanotechnology based approaches for enhancing oral bioavailability of poorly water-soluble antihypertensive drugs, *Scientifica* (Cairo). 2016. DOI: 10.1155/2016/8525679. 5

49.   Lin, M. H., Hung, C. F., Hsu, C. Y., Lin, Z.-C., and Fang, J.-Y. (2019). Prodrugs in combination with nanocarriers as a strategy for promoting antitumoral efficiency, *Future Med. Chem.* 11:2131–2150. DOI: 10.4155/fmc-2018-0388. 5

CHAPTER   2

Immune Response and Its Role in Cancer

Anubhab Mukherjee, Vijay Sagar Madamsetty, Sudip Mukherjee

2.1   ROLE OF THE BODY'S IMMUNITY IN REGULATION OF CANCER

It is interesting to decipher the enigmatic role that the immune system plays in extirpating the disease progression: incipient, late-stage, and micro-metastases of tumors. According to the well-established theory of immune surveillance, cells and tissues of the organisms are perpetually monitored by a wide-awake immune system, which can identify and thereby eradicate the bulk of incipient cancer cells. The proposition can be extrapolated to the fact that solid tumors must have somehow evaded immune destruction. A steady growth of a few cancers in immunocompromised patients vouched for this blemished immune surveillance of tumors. A bodacious accretion of evidence, from clinical epidemiology as well as from genetically engineered mice, hints that the immune system works as a weighty barrier to tumor formation and progression [1].

Compared to immunocompetent controls, tumors grew more often and more rapidly in genetically engineered immunodeficient mice. More particularly, a dearth of functional $CD8^+$ cytotoxic T lymphocytes (CTLs), $CD4^+$ T_h1 helper T cells, or natural killer (NK) cells, alone or together, resulted in significant enhancement of incidence of tumor. Furthermore, outcome of few transplantation experiments suggested that cancer cells originated from immunodeficient mice were almost always ineffective toward inducing secondary tumors in syngeneic immunocompetent hosts, whereas opposite results were obtained in a reverse tumor-host set up [2, 3]. Data from clinical epidemiology also fortify the presence of anti-tumor immunity in certain human cancers [4, 5]. For instance, a better prognosis is observed in patients with colon and ovarian cancer markedly infiltrated with killer lymphocytes when compared to those lacking in CTLs and NK cells [6, 7]. Moreover, post organ transplantation, immunosuppressed recipients developed donor-derived tumors—hinting at the possibility that the donors already contained these cancerous cells but they were under strict inspection of the immune system [8].

There is no other simpler way to convey that cancer cells with high immunogenicity may well escape immune annihilation by mutilating specific components of the immune system that have been deployed to demolish them. Cancer cells routinely incapacitate infiltrating cytotoxic lymphocytes (CTLs and NK cells) by exuding TGFβ or other immunosuppressive factors [9, 10]. Mechanistically, to accomplish this, immunosuppressive inflammatory cells, including regulatory T cells (T_{regs}) and myeloid-derived suppressor cells (MDSCs), are recruited as both are able to disable cytotoxic lymphocytes [11, 12]. The paradoxical co-existence of both tumor-promoting and tumor-antagonizing immune cells can be accounted for by contemplating the manifold functions of the immune system: an adaptive immune system specifically detects and eliminates infectious agents (supported by cells of the innate immune system) and the innate immune system is engaged in healing wounds and tissue housecleaning. The latter subtypes of immune cells (specifically, the alternatively activated macrophages), the key sources of the angiogenic, epithelial, stromal growth factors, and matrix-rebuilding enzymes required for wound healing, are engaged to support tumor progression.

2.2   INFLAMMATION AND CANCER: BIOLOGY OF IMMUNE RESPONSE IN CANCER THERAPY

Clinicians and pathologists have long identified that some tumors are densely infiltrated by cells of both the innate and adaptive arms of the immune system, giving rise to inflammatory conditions [13]. It was initially assumed and hypothesized that such immune responses were attempts by the immune system to eradicate tumors. As time rolled on, clues were being accumulated that the tumor-associated inflammatory response had the unforeseen, ironical effect of enhancing tumorigenesis and progression, in turn, helping incipient neoplasias to obtain salient features of cancer and fostering their eventual development into full-blown cancers [14–16]. Inflammation contributes to neoplastic growth by providing a growing list of signaling molecules: tumor growth factor EGF, the angiogenic growth factor VEGF, other proangiogenic factors such as FGF2, chemokines, and cytokines promoting the inflammatory state, proangiogenic and/or pro-invasive matrix-degrading enzymes, including MMP-9 and other matrix metalloproteinases, cysteine cathepsin proteases, heparanase, and inductive signals that lead to activation of EMT, etc. Additionally, inflammatory cells can release chemicals, notably reactive oxygen species (ROS), which are proactively mutagenic for neighboring cancer cells, accelerating their genetic evolution toward a state of heightened malignancy [17, 18].

2.3   MACROPHAGES AND THEIR ROLE IN CANCER IMMUNE RESPONSE

Growing knowledge of cancer immunology has led to the recognition that a cauldron of cells of the innate immune system generated in the bone marrow, such as—macrophages, neutrophils, mast cells, and myeloid progenitors—play pivotal roles in tumor angiogenesis [17, 19–20]. It is also clear that macrophages infiltrate incipient neoplasias as well as advanced tumors and gather at the periphery promoting local invasion and metastases by providing matrix-degrading enzymes, namely, metalloproteinases and cysteine cathepsin proteases [21, 22]. It has also been found that macrophage and cancer bear a vicious connection. In order to expedite tissue invasion and metastases, tumor-associated macrophages (TAMs) impart EGF to breast cancer cells with malignant propensity, while getting reciprocally triggered with colony stimulating factor-1 (CSF-1) by cancer cells [23, 17].

In truth, TAMs promote tumor growth and/or tumor immune suppression both through non-immune and immune pathways. On one hand, they secrete a high amount of VEGFs fueling neovascularization and enabling intravasation and dissemination of malignant cells [24]. On the other, they suppress of anticancer immune responses by reduced production of proinflammatory cytokine interleukin (IL)-12, production of immunosuppressive factors such as IL-10, TGFβ, and prostaglandin E2 (PGE2) to recruit immunosuppressive T regulatory cells (T_{REG}) via the chemokine CCL22 [25]. In addition, macrophages promote the extravasation of disseminated malignant cells at the metastatic site, form a supportive metastatic niche, aid to their survival by impeding immune clearance, and prompt tumors to trigger pro-survival signaling pathways [26].

It is, therefore, promising to develop agents (targeting macrophages in the tumor microenvironment) that can inhibit the recruitment of macrophages and/or hinder the tumorigenic effector functions of macrophages in both primary and metastatic tumor to ameliorate overall cancer survival. While a plethora of pre-clinical studies have distinguished key recruitment, polarization, and metabolism of TAMs during tumor progression, a vast number of clinical trials are also underway [27, 28].

2.4   CURRENT CANCER IMMUNOTHERAPY APPROACHES

As discussed in Section 2.1, tumors evade immune destruction by downmodulating the components of antigen processing and presentation to CTLs, by recruiting immune-suppressor T_{reg} cells, MDSCs, TAMs, etc., by secreting soluble immunosuppressive factors viz. TGF-β, IL-10, etc., and by upregulating ligands for co-inhibitory receptors that hinder the activity of tumor-infiltrating lymphocytes (TIL), viz. cytotoxic T lymphocyte antigen-4 (CTLA-4) and programmed death ligand-1 (PD-L1). These daunting challenges have precisely been addressed by current approaches of cancer immunotherapies, namely: (i) cytokines, (ii) co-stimulatory receptor agonists, (iii) checkpoint inhibitors, (iv) engineered T cells, (v) cancer vaccines, etc. Below, we shall briefly discuss the latter three of them.

2.4.1   IMMUNE CHECKPOINT BLOCKADE THERAPY

As we make it clear in our discussion, T cells play pivotal roles in adaptive anti-tumor immunity which is why the immense magnitude of scientific effort is directed toward inducing T-cell mediated anti-tumor responses. Of more than 20 co-stimulatory and co-inhibitory checkpoint molecule pairs available, a major breakthrough came from the identification of two pairs of inhibitory receptor/ligands and targeting them with antibodies subsequently: cytotoxic T lymphocyte-associated antigen-4 (CTLA-4) receptor with B7 ligand, and programmed cell death protein-1 (PD-1) receptor with PD-L1 ligand [29–32]. Mechanistically, it is worth mentioning here that CTLA-4 and PD-1 are co-inhibitory receptors expressed on the T cell surface which, when bound to their respective ligands (CD80/86 and PD-L1/-L2), render T cells anergic [33]. The tumor microenvironment allows the aberrant expression of immune checkpoint ligands (on both immune and tumor cells) which results in an untoward suppression of T-cell function [34]. A breach of such interactions can therefore restore the anti-tumor immune response from T cells, expanding their targets [35, 36]. This actually led to the USFDA approval of Ipilimumab (anti-CTLA-4 antibody, 2011) and Pembrolizumab (anti-PD-1 antibody, 2016) for the treatment of melanoma [37, 38].

2.4.2   CAR T CELL ADOPTIVE IMMUNOTHERAPY

Chimeric antigen receptor (CAR) T cells are genetically engineered T cells which express an artificial chimeric receptor, i.e., single-chain antibodies linked to a transmembrane region and an intracellular domain (having antigen-binding as well as T-cell activating capabilities) to target a specific protein—a tumor antigen [39, 40]. Precisely, the process involves—harvesting T cells from patients, transducing them with chimeric genes, then re-infusing the CAR-T cells into patients to recognize antibody's cognate antigen, and bind and proliferate until it reaches cytotoxicity [41]. Pioneering studies involving CD19 CAR-transduced human peripheral blood T lymphocytes showed improved anti-tumor efficacy to eradicate cancers of B-cell origin, namely acute lymphoblastic leukemia (ALL) and diffuse large B-cell lymphoma (DLBCL) [42, 43]. Recently, efforts are directed toward harnessing benefit from CAR-T cells having their target as CD30 for refractory Hodgkin's lymphoma, B-cell maturation antigen (BCMA) for multiple myeloma, CD33 for acute myeloid leukemia (AML), etc. [44]. With ~364 clinical trials ongoing worldwide, the first two of its kind that received FDA approval are Tisagenlecleucel (Kymriah/Novartis) for ALL and axicabtagene ciloleucel (Yescarta/Kite Pharma) for DLBCL [45].

2.4.3   THERAPEUTIC CANCER VACCINES AND TUMOR NEOANTIGENS

An utmost clarity of understanding of tumor microenvironment as well as immunosuppressive factors and identification of appropriate tumor-associated or specific antigens (TAA or TSA) are quintessential prerequisites for designing cancer vaccines. Various categories of cancer vaccines encompass tumor cell lysate, dendritic cells (DC), nucleic acids (DNA or mRNA), or neo-antigens, of which DC-based vaccines are most widely explored [46]. Briefly, the required steps to prepare DC vaccines are: DC collection from patient, compelling it by genetic engineering to express TAA, reinfusion into patient's body, antigen presentation, and thereby activation of T cells followed by cancer cell death [47]. In 2010, Sipuleucel-T, a DC-based vaccine (engineered by exposing them to prostatic acid phosphatase and GM-CSF) was approved by the FDA to combat metastatic castrate-resistant prostate cancer based on its ability to protract median survival by four months [48, 49].

DNA or mRNA-based vaccines also have drawn significant attention from researchers which involves uptake of exogenous nucleic acids by the antigen presenting cells (APC), antigen expression and presentation to T cells, and activation of T cells against tumor cells [50, 51]. Another interesting category of cancer vaccine, known as neoantigen vaccine, also has been well investigated in the last few years. These are TSAs arising from somatic DNA mutations in cancer cells. There are a few major advantages of neoantigen vaccines: (i) off-target adverse effects are eliminated owing to their specific presence in cancer cells; and (ii) multiple neoantigen can be used which is particularly useful to treat heterogeneous cancers [52, 53].

2.5   CHALLENGES AND RECENT ADVANCEMENTS IN ALTERNATIVE TREATMENT STRATEGIES

The above-mentioned five different classes of immunotherapy (three discussed here) inexorably face challenges related to the delivery of immunotherapeutics. The clinical success of the checkpoint inhibitors deeply depends on their interaction with the targeted receptors/proteins. Also, the key obstacle for their use is that they trigger considerable autoimmune response resulting in adverse effects which restrict the expected or allowed dose to be administered [54]. To diminish this untoward off-target effects, targeted delivery systems with the ability of controlled release are being designed and developed [55].

Considering the degree of infiltration (high or low) of the cytotoxic lymphocytes in the microenvironment, solid tumors can be classified into two groups: the one with high immunogenicity will manifest a stronger response to checkpoint inhibitors compared to the one with low immunogenicity [56]. Fairly, there is a chance here to explore the potential of delivery technologies to enhance immunogenicity in the latter group. Furthermore, owing to their potential to minimize systemic toxicity by reducing drug exposure to other tissues, combinatorial delivery approaches can be adapted, which otherwise would render high toxicity to the patients [57, 58].

Immunotherapeutics demanding an intracellular delivery (nucleic acid vaccines and small-molecule agonists) must circumvent extracellular and intracellular barriers in order to avoid exerting systemic toxicity. As both cell membrane and nucleic acids are negatively charged, DNA and RNA require a delivery vehicle, most suitably a lipid or polymer-based cell transfection agent. Moreover, DNA vaccines must cross the plasma and nuclear membranes for their transcription. On the other hand, mRNA-based vaccines, which are subject to quick degradation by nucleases, only need to be percolated into the cell cytosol for translation [59, 60]. Both of them required suitable and efficient delivery vehicles (preferably, nanomedicine) to overcome the inherent challenges and reach the expected achievement level. In the following chapters, we shall reflect upon various delivery techniques of immunotherapeutics.

2.6   REFERENCES

1.   Vajdic, C. M. and van Leeuwen, M. T. (2009). Cancer incidence and risk factors after solid organ transplantation, *Int. J. Cancer* 125:1747–1754. DOI: 10.1002/ijc.24439. 11

2.   Teng, M. W. L., Swann, J. B., Koebel, C. M., Schreiber, R. D., and Smyth, M. J. 2008 Immune-mediated dormancy: an equilibrium with cancer, *J. Leukoc. Biol.* 84:988–993. DOI: 10.1189/jlb.1107774. 11

3.   Kim, R., Emi, M., and Tanabe, K. (2007). Cancer immunoediting from immune surveillance to immune escape, *Immunology* 121:1–14. DOI: 10.1016/B978-012372551-6/50066-3. 11

4.   Bindea, G., Mlecnik, B., Fridman, W. H., Pagès F., and Galon, J. (2010). Natural immunity to cancer in humans, *Curr. Opin. Immunol.* 22:215–222. DOI: 10.1016/j.coi.2010.02.006. 11

5.   Ferrone, C., and Dranoff, G. (2010). Dual roles for immunity in gastrointestinal cancers, *J. Clin. Oncol.* 28:4045–4051. DOI: 10.1200/JCO.2010.27.9992. 11

6.   Nelson, B. H. (2008). The impact of T-cell immunity on ovarian cancer outcomes, *Immunol. Rev.* 222:101–116. DOI: 10.1111/j.1600-065X.2008.00614.x. 11

7.   Pagès F., Galon, J., Dieu-Nosjean, M. C., Tartour, E., Sautes-Fridman, C., and Fridman, W. H. (2010). Immune infiltration in human tumors: a prognostic factor that should not be ignored, *Oncogene* 29:1093–1102. DOI: 10.1038/onc.2009.416. 11

8.   Strauss, D. C. and Thomas, J. M. (2010). Transmission of donor melanoma by organ transplantation, *Lancet Oncol.* 11:790–796. DOI: 10.1016/S1470-2045(10)70024-3. 11

9.   Yang, L., Pang, Y., and Moses, H. L. (2010). TGF-beta and immune cells: an important regulatory axis in the tumor microenvironment and progression, *Trends Immunol.* 31:220–227. DOI: 10.1016/j.it.2010.04.002. 12

10.   Shields, J. D., Kourtis, I. C., Tomei, A. A., Roberts, J. M., and Swartz, M. A. (2010). Induction of lymphoidlike stroma and immune escape by tumors that express the chemokine CCL21, *Science* 328:749–752. DOI: 10.1126/science.1185837. 12

11.   Mougiakakos, D., Choudhury, A., Lladser, A., Kiessling, R., and Johansson, C. C. (2010). Regulatory T cells in cancer, *Adv. Cancer Res.* 107:57–117. DOI: 10.1016/S0065-230X(10)07003-X. 12

12.   Ostrand-Rosenberg, S. and Sinha, P. (2009). Myeloid-derived suppressor cells: linking inflammation and cancer, *J. Immunol.* 182:4499–4506. DOI: 10.4049/jimmunol.0802740. 12

13.   Dvorak, H. F. (1986). Tumors: wounds that do not heal. Similarities between tumor stroma generation and wound healing, *N. Engl. J. Med.* 315:1650–1659. DOI: 10.1056/NEJM198612253152606. 12

14.   DeNardo, D. G., Andreu, P., and Coussens, L. M. (2010). Interactions between lymphocytes and myeloid cells regulate pro- versus anti-tumor immunity. *Cancer Metastasis Rev.* 29:309–316. DOI: 10.1007/s10555-010-9223-6. 12

15.   de Visser, K. E., Eichten, A., and Coussens, L. M. (2006). Paradoxical roles of the immune system during cancer development, *Nat. Rev. Cancer* 6 24–37. DOI: 10.1038/nrc1782. 12

16.   Grivennikov, S. I., Greten, F. R., and Karin, M. (2010). Immunity, inflammation, and cancer., *Cell* 140:883–899. DOI: 10.1016/j.cell.2010.01.025. 12

17.   Qian, B. Z. and Pollard, J. W. (2010). Macrophage diversity enhances tumor progression and metastasis, *Cell* 141:39–51. DOI: 10.1016/j.cell.2010.03.014. 12, 13

18.   Colotta, F., Allavena, P., Sica, A., Garlanda, C., and Mantovani, A. (2009). Cancer-related inflammation, the seventh hallmark of cancer: links to genetic instability, *Carcinogenesis* 30:1073–1081. DOI: 10.1093/carcin/bgp127. 12

19.   Zumsteg, A. and Christofori, G. (2009). Corrupt policemen: inflammatory cells promote tumor angiogenesis, *Curr. Opin. Oncol.* 21 60–70. DOI: 10.1097/CCO.0b013e32831bed7e. 13

20.   Murdoch, C., Muthana, M., Coffelt, S. B., and Lewis, C. E. (2008). The role of myeloid cells in the promotion of tumor angiogenesis, *Nat. Rev. Cancer* 8:618–631. DOI: 10.1038/nrc2444. 13

21.   Ferrara, N. (2010). Pathways mediating VEGF-independent tumor angiogenesis., *Cytokine Growth Factor Rev.* 21:21–26. DOI: 10.1016/j.cytogfr.2009.11.003. 13

22.   Kessenbrock, K., Plaks, V., and Werb, Z. (2010). Matrix metalloproteinases: Regulators of the tumor microenvironment, *Cell* 141:52–67. DOI: 10.1016/j.cell.2010.03.015. 13

23.   Wyckoff, J. B., Wang, Y., Lin, E. Y., Li, J. F., Goswami, S., Stanley, E. R., Segall, J. E., Pollard, J. W., and Condeelis, J. (2007). Direct visualization of macrophage-assisted tumor cell intravasation in mammary tumors, *Cancer Res.* 67:2649–2656. DOI: 10.1158/0008-5472.CAN-06-1823. 13

24.   Kitamura, T., Qian, B.-Z., and Pollard, J. W., (2015). Immune cell promotion of metastasis, *Nat. Rev. Immunol.* 15:73–86. DOI: 10.1038/nri3789. 13

25.   Ostuni, R., Kratochvill, F., Murray, P. J., and Natoli, G. (2015). Macrophages and cancer: from mechanisms to therapeutic implications, *Trends Immunol.* 36(4) 229-39. DOI: 10.1016/j.it.2015.02.004. Epub 2015 Mar 11. PMID: 25770924. 13

26.   Nielsen, S. R. and Schmid, M. C. (2017). Macrophages as key drivers of cancer progression and metastasis, *Mediat. of Inflamm.*, vol. 2017, Article ID 9624760, 11 pages, DOI: 10.1155/2017/9624760. 13

27.   Pathria, P., Louis, T. L., and Varner, J. A. (2019). Targeting tumor-associated macrophages in cancer, *Trends Immunol.* 40(4):310–327. DOI: 10.1016/j.it.2019.02.003. Epub 2019 Mar 17. PMID: 30890304. 13

28.   Lin, Y., Xu, J., and Lan, H. (2019). Tumor-associated macrophages in tumor metastasis: biological roles and clinical therapeutic applications, *J. Hematol. Oncol.* 12:76. DOI: 10.1186/s13045-019-0760-3. 13

29.   Hodi, F. S., O'Day, S. J., McDermott, D. F., Weber, R. W., Sosman, J. A., Haanen, J. B., Gonzalez, R., Robert, C., Schadendorf, D., Hassel, J. C., Akerley, W., van den Eertwegh, A. J. M., Lutzky, J., Lorigan, P., Vaubel, J. M., Linette, G. P., Hogg, D., Ottensmeier, C. H., Lebbé, C., Peschel, C., Quirt, I., Clark, J. I., Wolchok, J. D., Weber, J. S., Tian, J., Yellin, M. J., Nichol, G. M., Hoos, A., and Urba, W. J. (2010). Improved survival with ipilimumab in patients with metastatic melanoma, *N. Engl. J. Med.* 363:711–723. DOI: 10.1056/ NEJMoa1003466. 14

30.   Larkin, J., Hodi, F. S., and Wolchok, J. D. (2015). Combined nivolumab and ipilimumab or monotherapy in untreated melanoma, *N. Engl. J. Med.* 373:1270–1271. DOI: 10.1056/ NEJMc1509660. 14

31.   Brunet, J.-F., Denizot, F., Luciani, M.-F., Roux-Dosseto, M., Suzan, M., Mattei, M.-G., and Golstein, P. (1987). A new member of the immunoglobulin superfamily–CTLA-4, *Nature* 328:267–270. DOI: 10.1038/328267a0. 14

32. Ishida, Y., Agata, Y., Shibahara, K., and Honjo, T. (1992). Induced expression of PD-1, a novel member of the immunoglobulin gene superfamily, upon programmed cell death, *EMBO J.* 11:3887–3895. DOI: 10.1002/j.1460-2075.1992.tb05481.x. 14

33. Schwartz, R. H. (2003). T cell anergy, *Annu. Rev. Immunol.* 21:305–334. DOI: 10.1146/annurev.immunol.21.120601.141110. 14

34. Schietinger, A. and Greenberg, P. D. (2014). Tolerance and exhaustion: defining mechanisms of T cell dysfunction, *Trends Immunol.* 35:51–60. DOI: 10.1016/j.it.2013.10.001. 14

35. Kvistborg, P., Philips, D., Kelderman, S., Hageman, L., Ottensmeier, D., Joseph-Pietras, D., Welters, M. J. P., van der Burg, S., Kapiteijm, E., Michielin, O., Romano, E., Linnemann, C., Speiser, D., Blank, C., Haanen, J. B., and Schumacher, T. N. (2014). Anti-CTLA-4 therapy broadens the melanoma-reactive CD8þ T cell response, *Sci. Transl. Med.* 6:254ra128. DOI: 10.1126/scitranslmed.3008918. 14

36. Ribas, A., Shin, D. S., Zaretsky, J., Frederiksen, J., Cornish, A., Avramis, E., Seja, E., Kivork, C., Siebert, J., Kaplan-Lefko, P., Wang, X., Chmielowski, B., Glaspy, J. A., Tumeh, P. C., Chodon, T., Pe'er, D., and Comin-Anduix, B. (2016). PD-1 blockade expands intratumoral T memory cells, *Cancer Immunol. Res.* 4:194–203. DOI: 10.1158/2326-6066. CIR-15-0210. 14

37. Mansh, M. (2011). Ipilimumab and cancer immunotherapy: a new hope for advanced stage melanoma, *Yale J. Biol. Med.* 84381-9. 14

38. Nishino, M., Ramaiya, N. H., Hatabu, H., and Hodi, F. S. (2017). Monitoring immune-checkpoint blockade: response evaluation and biomarker development, *Nat. Rev. Clin. Oncol.* 14:655–668. DOI: 10.1038/nrclinonc.2017.88. 14

39. Srivastava, S. and Riddell, S. R. (2015). Engineering CAR-T cells: Design concepts, *Trends Immunol..* 36(8):494–502. DOI: 10.1016/j.it.2015.06.004. 14

40. Sadelain, M., Brentjens, R., and Rivière, I. (2013). The basic principles of chimeric antigen receptor design, *Cancer Discovery.* 3(4):388–98. DOI: 10.1158/2159-8290.CD-12-0548. 14

41. Hartmann, J., Schüßler-Lenz, M., Bondanza, A., and Buchholz, C. J. (2017). Clinical development of CAR T cells-challenges and opportunities in translating innovative treatment concepts, *EMBO Molec. Med.* 9(9):1183–1197. DOI: 10.15252/emmm.201607485. 14

42. Kochenderfer, J. N., Wilson, W. H., Janik, J. E., Dudley, M. E., Stetler-Stevenson, M., Feldman, S. A., Maric, I., Raffeld, M., Nathan, D.-A., Lanier, B. J., Morgan, R. A., and

Rosenberg, S. A. (2010). Eradication of B-lineage cells and regression of lymphoma in a patient treated with autologous T cells genetically engineered to recognize CD19, *Blood* 116(20):4099–102. DOI: 10.1182/blood-2010-04-281931. 14

43.   Kochenderfer, J. N., Dudley, M. E., Kassim, S. H., Somerville, R. P. T., Carpenter, R. O., Stetler-Stevenson, M., Yang, J. C., Phan, G. Q., Hughes, M. S., Sherry, R. M., Raffeld, M., Feldman, S., Lu, L., Li, Y. F., Ngo, L. T., Goy, A., Feldman, T., Spaner, D. E., Wang, M. I., Chen, C. C., Kranick, S. M., Nath, A., Nathan, D.-A. N., Morton, K. E., Toomey, M. A., and Rosenberg, S. A. (2015). Chemotherapy-refractory diffuse large B-cell lymphoma and indolent B-cell malignancies can be effectively treated with autologous T cells expressing an anti-CD19 chimeric antigen receptor, *J. Clin. Oncol.* (33):540–9. DOI: 10.1200/JCO.2014.56.2025. 14

44.   Schultz. L. and Mackall, C. (2019). Driving CAR T cell translation forward, *Sci. Trans. Med.* 11(481):eaaw2127. DOI: 10.1126/scitranslmed.aaw2127. 14

45.   Yu, J. X., Hubbard-Lucey, V. M., and Tang, J. (2019-05-30). The global pipeline of cell therapies for cancer, *Nat. Rev. Drug Disc.* 18 (11) 821–822. DOI: 10.1038/d41573-019-00090-z. 14

46.   Guo, C., Manjili, M. H., Subjeck, J. R., Sarkar, D., Fisher, P. B., and Wang, X.-Y. (2013). Therapeutic cancer vaccines; past, present and future, *Adv. Cancer Res.* 119:421–475. DOI: 10.1016/B978-0-12-407190-2.00007-1. 15

47.   Garg, A. D., Coulie, P. G., Van den Eynde, B. J., and Agostinis, P. (2017). Integrating next- generation dendritic cell vaccines into the current cancer immunotherapy landscape, *Trends Immunol.* 38:577–593. DOI: 10.1016/j.it.2017.05.006. 15

48.   Kantoff, P. W., Higano, C. S., Shore, N. D., Berger, E. R., Small, E. J., Penson, D. F., Redfern, C. H., Ferrari, A. C., Dreicer, R., Sims, R. B., Xu, Y., Frohlich, M. W., Schellhammer, P. F., and IMPACT Study Investigators. (2010). Sipuleucel- T immunotherapy for castration- resistant prostate cancer, *N. Engl. J. Med.* 363:411–422. DOI: 10.1056/NEJMoa1001294. 15

49.   Mullard, A. (2016). The cancer vaccine resurgence, *Nat. Rev. Drug Discov.* 15:663–665. DOI: 10.1038/nrd.2016.201. 15

50.   Pardi N., Hogan, M. J., Porter, F. W., and Weissman, D. mRNA vaccines—a new era in vaccinology, *Nat. Rev. Drug Discov.* 17:261–279 (2018). DOI: 10.1038/nrd.2017.243. 15

51.   Yang, B., Jeang, J., Yang, A., Wu, T. C., and Hung C.-F. (2015). DNA vaccine for cancer immunotherapy, *Hum. Vaccin. Immunother.* 10:3153–3164. DOI: 10.4161/21645515.2014.980686. 15

52.   Li, L., Goedegebuure, S. P., and Gillanders, W. E. (2017). Preclinical and clinical development of neoantigen vaccines, *Ann. Oncol.* 28 xii11–xii17. DOI: 10.1093/annonc/mdx681. 15

53.   Lauss, M., Donia, M., Harbst, K., Andersen, R., Mitra, S., Rosengren, F., Salim, M., Vallon-Christersson, J., Törngren, T., Kvist, A., Ringnér, M., Svane, I. M., and Jönsson, G. (2017). Mutational and putative neoantigen load predict clinical benefit of adoptive T cell therapy in melanoma, *Nat. Commun.* 8:1738. DOI: 10.1038/s41467-017-01460-0. 15

54.   June, C. H., Warshauer, J. T., and Bluestone, J. A. (2017). Is autoimmunity the Achilles' heel of cancer immunotherapy?, *Nat. Med.* 23:540–547. DOI: 10.1038/nm.4321. 15

55.   Riley, R. S., June, C. H., Langer, R., and Mitchell, M. J. (2019). Delivery technologies for cancer immunotherapy, *Nat. Rev. Drug. Discov.* 18:175–196, DOI: 10.1038/s41573-018-0006-z. 15

56.   Binnewies, M., Roberts, E. W., Kersten, K., Chan, V., Fearon, D. F., Merad, M., Coussens, L. M., Gabrilovich, D. I., Ostrand-Rosenberg, S., Hedrick, C. C., Vonderheide, R. H., Pittet, M. J., Jain, R. K., Zou, W., Howcroft, T. K., Woodhouse, E. C., Weinberg, R. A., and Krummel, M. F. (2018). Understanding the tumor immune microenvironment (TIME) for effective therapy, *Nat. Med.* 24:541–550. DOI: 10.1038/s41591-018-0014-x. 15

57.   Milling, L., Zhang, Y., and Irvine, D. J. (2017). Delivering safer immunotherapies for cancer, *Adv. Drug Deliv. Rev.* 114:79–101. DOI: 10.1016/j.addr.2017.05.011. 15

58.   Song, W., Shen, L., Wang, Y., Liu, Q., Goodwin, T. J., Li, J., Dorosheva, O., Liu, T., Liu, R., and Huang, L. (2018). Synergistic and low adverse effect cancer immunotherapy by immunogenic chemotherapy and locally expressed PD- L1 trap, *Nat. Commun.* 9:2237. DOI: 10.1038/s41467-018-04605-x. 15

59.   McNamara, M. A., Nair, S. K., and Holl, E. K. (2015). RNA- based vaccines in cancer immunotherapy, *J. Immunol. Res.* 794528. DOI: 10.1155/2015/794528. 16

60.   Zhu, G., Zhang, F., Ni, Q., Niu, G., and Chen, X. (2017). Efficient nanovaccine delivery in cancer immunotherapy, *ACS Nano.* 11:2387–2392. DOI: 10.1021/acsnano.7b00978. 16

CHAPTER 3

Immunomodulation and Various Strategies Effecting Immune Response

Sudip Mukherjee, Anubhab Mukherjee, Vijay Sagar Madamsetty

3.1 IMMUNE RESPONSE TO THE IMPLANTED BIOMATERIAL

External biomaterials/nanomaterials introduced in the body effect in an interruption of the local or systemic host tissue environment [1, 2]. In the initial process of implantation, injury to the vascularized connective tissue stimulates the preliminary inflammatory reaction (innate immunity) that instigates blood–biomaterial interactions. Plasma components and proteins from blood exudates are adsorbed on the surface of biomaterials within a few minutes and adsorption of blood proteins and platelet activation results in recruiting and activation of neutrophils (polymorphonuclear leukocytes) and mast cells, which characterize the acute inflammation phase [3–5]. The acute inflammatory response with biomaterials generally resolves rapidly, usually less than one week, depending on the extent of the injury at the implant site. Degranulation of mast cells with histamine release and fibrinogen adsorption mediates the acute inflammatory responses to implanted biomaterials [6, 7]. Monocytes are stimulated in the transition from acute to chronic inflammation. Monocytes migrate to the wound site and differentiate into macrophages [8, 9]. The release of several pro-inflammatory cytokines, chemokines that induce directed chemotaxis of other innate inflammatory cells, and maturation of dendritic cells leads to the activation of adaptive immunity via B and T lymphocytes [10]. During the initial phase of chronic inflammation, there are an increase in specific pro-inflammatory (M1) macrophages in the injured site [11]. The M1 subtype secretes numerous enzymes, including collagenases and different cytokines, such as TNF-α, IL-1, IL-6, IL-8, and IL-10, which further stimulate the inflammatory response [10]. Based on stimulation by T_H lymphocytes, the polarization of the macrophage population can change to a pro-healing phenotype (M2), which can be stimulated by several factors such as glucocorticoids, IL-4, or IL-13. These factors release anti-inflammatory cytokines to resolve inflammation [12]. When this switch is made, tissue cells are stimulated toward the regeneration of the implant site. However, if the environment fails to induce the switch to an abundance of M2 macrophages, the chronic inflammation phase will lead to fibrosis [10]. Macrophages undergo fusion to improve their efficiency and form multinucleated giant cells, known as foreign-body giant cells (FBGCs) [3]. These FBGCs send out signals for fibroblasts to arrive at the injury site. The implant is effectively isolated from the rest of the body because of collagen deposition by the fibroblasts. This condition, in which the fibrous capsule encapsulates the implant, along with the interfacial foreign body reaction, is known as fibrosis [13].

3.2   IMMUNOMODULATION STRATEGIES

As previously described, implanted or injected materials induce a host reaction leading to a FBR that regulates the conclusion of the integration and the biological functioning of the implant [14]. These cascades of inflammatory responses prompt wide biochemical signals that recruits various immune cells near the newly introduced materials [12, 15]. Importantly, the physicochemical properties of these introduced materials perform a substantial role in macrophage plasticity, causing the tissue healing process. Hence, material scientists have established various immunomodulatory materials to tune appropriate immune responses as per their requirements. In this chapter three main strategies will be reviewed for tuning inflammatory response including: (1) delivery of biological molecules or cells; (2) modification of the chemical properties; and (3) alteration of physical properties of the materials.

3.2.1   IMMUNOMODULATION VIA BIOACTIVE MOLECULES/CELL DELIVERY

Delivering biologics including anti-inflammatory drugs or cytokines using external nanomaterials/biomaterials has achieved immense attention as an approach to regulate cross-talk among engaged immune cells and implant adjacent tissues. Cytokines and growth factors dynamically regulate the phenotype of the immune cells, along with sustained release from bioactive coatings [10]. Cytokines can be locally delivered either by immobilization into the biomaterial or by nucleic-acid-based strategies that allow prolonged cytokine production from in situ implanted engineered cells [14]. For example, the hydrogel inclusion of TGF-β or IL-10 is effective in suppressing the maturation of dendritic cells [16]. The sequential controlled delivery of IFN-γ and IL-4 from scaffolds or double hydrogel layers promoted the transition of M1 to M2 macrophages [14, 17, 18]. Also, the delivery of soluble pharmacological anti-inflammatory agents, such as heparin [19], dexamethasone [20], and superoxide dismutase [21] from reservoirs and coatings, has shown reduced inflammation and fibrous encapsulation [10]. Swartzlander et al. reported that encapsulated MSCs attenuated the fibrotic response of the FBR compared to acellular hydrogels by downregulating the classically activated macrophages [22]. In another study, the conditioned medium obtained from macrophages induced by specific biomaterials could differentiate other cells [23].

3.2.2   IMMUNOMODULATION STRATEGIES USING BIOMATERIAL CHEMISTRY

Biomaterials surfaces can be modified with various types of chemistry for the modulation of protein adsorption and consequent cellular behavior. It is notable to mention that alteration in the surface chemistry by tuning surface properties can influence the biological responses of immune-associated factors. Various types of chemical approaches were taken to mitigate the FBR and prevent fibrosis.

Use of Non-Biofouling Strategies

Significant developments were created to mitigate FBR using immune isolating materials with a coated implant surface [24, 25]. Non-fouling thick hydrophilic polymeric brushes and films can passively decrease the protein adsorption and FBR [24, 25]. Semi-porous hydrogels have been extensively utilized as anti-fibrotic coatings due to their exclusive properties such as excess water content, accessible transport of solute, and easy surface functionalization strategies by additional chemical modification [24, 25]. The use of poly(N-isopropyl acrylamide) (pNIPAm) films cross-linked with poly(ethylene glycol) (PEG) was used for prevention of any primary human macrophage adhesion *in vitro* to mitigate FBR. It further attenuates a severe leukocyte adhesion and the amount of pro-inflammatory cytokines *in vivo*. Recently, more active strategies were also employed by various authors by forming self-healing spongy surfaces [26].

Use of Bioinspired Extracellular Matrix (ECM) Components

Immunomodulation using novel biomaterials that use ECM components and mimic natural microenvironment can induce natural tissue regeneration or wound healing [27]. A recent report by Kajahn et al. showed highly sulfated hyaluronan (HA) fabricated materials can prevent IFN-g-, IL-6-, and MCP-1 mediated M1-activation of macrophages which is the reason behind improper wound healing around implants [28]. A similar way Brown et al. reported that the composition of biomaterials implant plays a vital role in controlling the M2:M1 macrophage ratio around the implants ultimately regulating wound healing [29].

Use of Chemically Modified Hydrogels for Immunoisolation of Bioactive Cells

Recent advancements in cell therapy have gained huge attention due to potential advantages over chemical drugs. The transplantation technologies using donor cells require long term protection from FBR and host recognition. Also, a semi-permeable hydrogel network is needed to allow small molecules, such as NO, ROS, O_2, and cytokines, to effortlessly circulate while avoiding close contact between encapsulated cells and infiltrating immune cells around these implants. Lin et al. used the RGD-functionalized PEG hydrogels and further modified with peptides that show strong binding affinity to these pro-inflammatory cytokines, including TNF-α and MCP-1, causing better cells survival compared to unmodified PEG hydrogels [30].

Hydrogel's surfaces were modified by immunomodulatory small molecules to prevent any FBR. Vegas and Veiseh et al. demonstrated a unique chemical modification strategy in alginate-based hydrogels using small molecules for the modulation of host immune-recognition, reason for FBR [31, 32]. A varied set containing 774 small molecule conjugated alginate analogs were synthesized using a combinatorial approach. To monitor the early inflammation (an early indicator of immune recognition) and *in vivo* screening approach were carried out measuring the cathepsin activity (a marker for degranulation) seven days after injecting the small molecule modified alginate-based hydrogels subcutaneously. This reduced the lead to an initial 200 with reduced cathepsin activity compared to control of unmodified alginate hydrogels. An additional screening narrowed down the final 10 lead alginate analogs with the lowest inflammation. Intraperitoneal (IP) implant using lead materials for 14 days in mice established lowest FBR, cell, macrophage, and collagen deposition for 3 lead analogs which were confirmed by phase-contrast and immunofluorescence staining (Figure 3.1). Further studies confirmed notable reductions in collagen, α-smooth muscle actin, macrophages, and neutrophils extracted from the retrieved hydrogels. Intravital imaging demonstrated a remarkable decrease in macrophage cell density around small molecule modified alginate confirming the lowest host recognition and consequent least FBR. 1.5 mm capsules of the three top materials were implanted into the IP space of non-human primates (NHP, cynomolgus macaques) along with unmodified alginate capsules as a control, exhibited similar results like rodents with minimal FBR. Mechanistic studies (including the mechanical strength, protein adsorption, and surface roughness, of the chemically modified alginates compared to unmodified alginates) did not disclose any other potential cause for the difference in FBR confirming the direct relation of FBR with chemical modification. It is essential to highlight that all three top materials had triazole framework in their chemical structure that may play an essential role in mitigating fibrosis. Related studies confirmed a similar result [33].

Use of Zwitterionic Biomaterials for Immunomodulation

Low fouling zwitterionic biomaterials showed potent prevention against FBR. Jiang et al. implanted PCBMA hydrogels subcutaneously in the C57BL/6 mice followed by observation of FBR and inflammatory response over time showed minimal collagen deposition compared to pHEMA, even three months post-implantation [34]. The mechanistic studies revealed the growth of angiogenic blood vessels around the PCBMA implants, along with the presence of a considerable number of macrophages, showed anti-inflammatory expressions compared to pHEMA. Anderson et al. employed a similar strategy to modify alginate microspheres coated with a zwitterionic poly (methacryloyloxyethyl) phosphorylcholine (pMPC) containing copolymer using a dopamine-based conjugation strategy, showing lower FBR in mice [35]. A similar strategy was further utilized to improve the function of continuous blood glucose monitors (CGM) [36]. A total of 64 diverse zwitterionic hydrogels were used to coat the sensors, and the efficacies of the devices were tested *in vivo*. Results showed that Poly(MPC)-coated sensors observed the lowest inflammation, FBR, and better CGM applications *in vivo*.

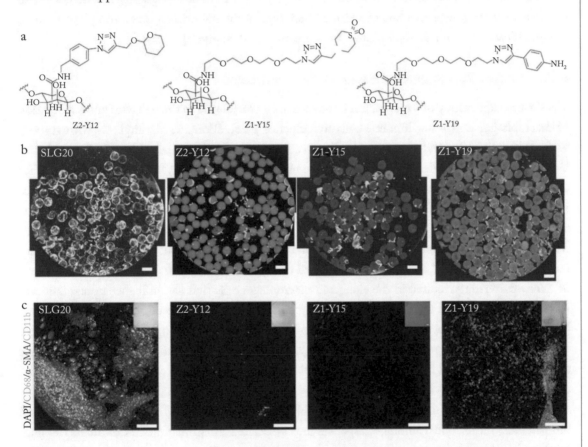

Figure 3.1: Small-molecule modifications of the alginate backbone can mitigate the FBR. (a) The lead small-molecule modifications that were identified from a combinatorial modified alginate library. These lead modified alginates successfully mitigated fibrosis in the IP space of both immunocompetent murine and NHP animal models. (b) Representative phase-contrast imaging of retrieved capsules for the control (SLG20) and modified alginate formulations after four weeks inNHP are shown. (c) Immunofluorescence imaging of the retrieved capsules in (B) showed reduced markers of general cellular material (DAPI), myofibroblasts (α-SMA), and macrophages (CD11b). The figure and caption are reproduced after permission from Spring Nature, 2016 [31].

3.2.3   IMMUNOMODULATION USING PHYSICAL BIOMATERIALS PROPERTIES

Biomaterial's physical properties, including size, shape, surface morphology, roughness, topography, and geometry, play a crucial role in orchestrating protein adsorption, macrophage attachment, and host immune responses to foreign materials [13, 14, 37]. Material topography is well known to manipulate macrophage attachment and the phenotype of the material's surface also plays a role in the modification of an adsorbed protein by conformational changes [37, 38].

Role of Surface Topography in Immunomodulation (Texture)

The surface topography of a biomedical implant plays a crucial role in the behavior and modulation of macrophages and other immune cells to influence FBR. Biomaterials implants with smooth surfaces usually show less macrophage adhesion, low level of inflammation, less FBR, and better biocompatibility [14]. Manipulating the surface nano/microtopography, hydrophilicity, and surface charge can influence host immunological responses to biomaterials. Moreover, alteration in surface roughness at the nanoscale can influence higher protein adsorption, affecting interactions with immune cells [39, 40], and different nanostructured topographies can affect cellular interactions [41]. McWhorter et al. demonstrated the modulation of micro and nano-topographical features to determine macrophage cell attachment, shapes, and orientation. The authors showed that modulation of micro-patterning resulted in the elongated macrophage cells that can influence upregulation of M2-markers associated with mouse macrophage and a notable decrease in proinflammatory cytokine secretion and FBR [42]. Studies revealed that imprinting the grating to the polymer surfaces caused behavioral changes of macrophages independent of grating size or surface chemistry. Larger size gratings imprinted on polymer surfaces influenced the adhesion of immune cells than planar polymeric control surfaces.

Although titanium surfaces are considerably more biocompatible than most of the implanted biomaterials, more robust immune responses and associated inflammation can result in titanium implant failure. Several research groups have altered the surface topography using micro-nano engineering to reduce the immune responses. Moreover, researchers have modulated the surface hydrophobicity and roughness to reduce the immune response and potential FBR. This has led to research focused on possible surface modifications to decrease the immune response in titanium structures. This can be achieved by several methods, such as nano/microstructures [43]. Alfarsi et al. demonstrated a significant decrease in pro-inflammatory genes to the hydrophilic modified sandblasted, acid-etched (SLA) implants when compared to control SLA surfaces [44]. For further details, a recent review by Rostam et al. can be further followed, where the authors have shown the impact of biomaterial surface topography and chemistry on the function of antigen-presenting cells (APCs) [45]. Vassey et al. recently reported a high throughput screening approach that showed micropillars of 5–10 µm diameter participate in a leading role in macrophage attachment [37]. Combining the micropillar density with the micropillar size was identified as central for the modulation of cell phenotypes from pro- to anti-inflammatory states.

Role of Implant/Biomaterials Geometry in Immunomodulation (Size and Shape)

Various studies showed the role of the geometry of biomaterials implants, including size and shape, for controlling FBR [43, 46, 47]. Ward et al. demonstrated micro cylinders, made from various materials, exhibited a different range of FBR depending on the cylinder diameters, the porosity of the materials after subcutaneous implantation for seven weeks in a Sprague-Dawley rat. The authors found that thin cylinders (0.3 mm) compared to thick cylinders (2.0 mm) had low FBR and a thinner capsule [48].

In a pioneering work by Salthouse et al. demonstrated *in vivo* biocompatibility of rods made from different medical-grade materials with varying cross-sections [49]. The authors showed that rods with circular cross-sections elicited the lowest FBR than triangular and pentagonal cross-sections. Other important studies demonstrated that smooth surface materials produce less acute immune reactions compared to implants with corners, sharp features, and acute angles [50]. Veiseh et al. further reported the function of implant geometry, including the size of implants on their long-term *in vivo* biocompatibility [50]. They implanted different spherical implants sized from 0.3–1.9 mm in diameter in rodents and primates and showed that larger implants of 1.5-mm diameter and above elicit very low FBR for an extended period *in vivo* (Figure 3.2). Similar observations were recorded using different sets of materials, including alginate hydrogels, glass, metals, and plastics. The effect was independent of the total surrounded surface area. Other published reports showed that the size of titanium nanotubes can be modulated to reduce macrophage attachment, and FBR (60–70 nm nanotube showed lowest macrophage attachment) [51].

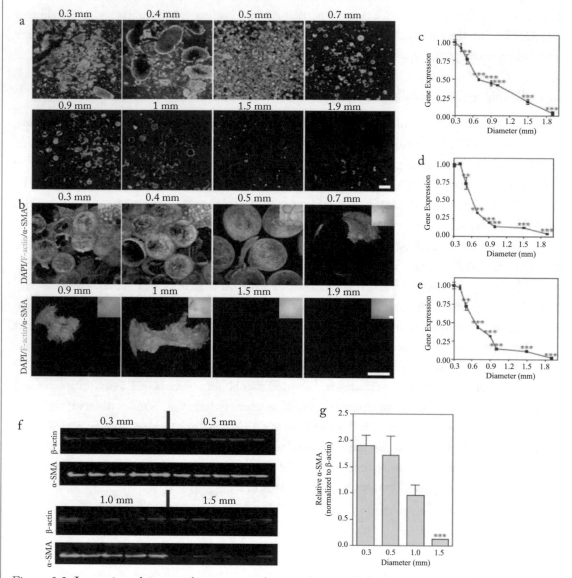

Figure 3.2: Increasing alginate sphere size results in reduced cellular deposition and fibrosis formation on the spheres. SLG20 alginate spheres (0.5 mL 30 in volume) of eight different sizes (0.3, 0.4, 0.5, 0.6, 0.7, 1, 1.5, and 1.9 mm) were implanted into the intraperitoneal space of C57BL/6 mice, where they were retained for 14 days and analyzed for degree of fibrosis on retrieval. (a) Dark-field phase-contrast images obtained from retrieved spheres reveal a significant decrease in the level of cellular overgrowth with an increase in sphere size. Scale bar, 2 mm. (b) Z-stacked confocal images of retrieved spheres immunofluorescence stained with DAPI (highlighting cellular nuclei), phalloidin (highlighting F-actin), and α-SMA (highlighting myofibroblast cells). Scale bar, 300 m. (c-e), qPCR-based expression analysis of fibrotic markers α-SMA (c), Collagen 1a1 (d, Col1a1) and Collagen 1a2 (e, Col1a2) directly on the

eight different sphere sizes (0.3, 0.4, 0.5, 0.7, 0.9, 1, 1.5, and 1.9 mm) plotted normalized to relative expression levels on 300-m spheres. (f) Semiquantitative western blot analysis of α-SMA expression in cell overgrowth on microspheres. (g) Plot of analyzed band intensities from western blot images shown in (f). The figure and caption are reproduced after permission from Spring Nature, 2015 [50].

3.3 CLASSIFICATIONS OF IMMUNOMODULATORY NANOMEDICINE

Immunomodulatory materials can be present in form of either small nanomaterials-based form or as a larger biomaterials-based implants (capsules, blocks, 3-D printed hydrogels, discs, etc). In the present book, we will mostly discuss the various types of immunomodulatory nanomaterials that have roles in cancer therapeutic applications including: (a) organic and polymer nanomedicine; (b) liposomes and lipid nanoparticles; (c) protein nanoparticles; and (d) viral nanoparticles.

3.4 REFERENCES

1.   Grainger, D. W. (2013). All charged up about implanted biomaterials, *Nat. Biotechnol.*. 31(6):507-9. DOI: 10.1038/nbt.2600. 23

2.   Williams, D. F. (2008). On the mechanisms of biocompatibility, *Biomaterials* 29(20):2941-53. DOI: 10.1016/j.biomaterials.2008.04.023. 23

3.   Sheikh, Z., Brooks, P. J., Barzilay, O., Fine, N., and Glogauer, M. (2015). Macrophages, foreign body giant cells and their response to implantable biomaterials, *Materials* (Basel). 8(9):5671-701. DOI: 10.3390/ma8095269. 23

4.   Lock, A., Cornish, J., and Musson, D. S. (2019). The role of *in vitro* immune response assessment for biomaterials, *J. Funct. Biomater.* 10(3):31. DOI: 10.3390/jfb10030031. 23

5.   Kolaczkowska, E. and Kubes, P. (2013). Neutrophil recruitment and function in health and inflammation., *Nat Rev Immunol* 13(3):159-75. DOI: 10.1038/nri3399. 23

6.   Anderson, J. M., Rodriguez, A., and Chang, D. T. (2008). Foreign body reaction to bio-materials, *Sem. Immunol.* 20(2):86-100. DOI: 10.1016/j.smim.2007.11.004. 23

7.   Luster, A. D., Alon, R., and von Andrian, U. H. (2005). Immune cell migration in in-flammation: present and future therapeutic targets, *Nat. Immunol.* 6(12):1182-90. DOI: 10.1038/ni1275. 23

8.   Gordon, S. and Taylor, P. R. (2005). Monocyte and macrophage heterogeneity, *Nat Rev Immunol* 5(12):953-64. DOI: 10.1038/nri1733. 23

9.   Soehnlein, O. and Lindbom, L. (2010). Phagocyte partnership during the onset and res-
        olution of inflammation, *Nat. Rev. Immunol.* 10(6):427-39. DOI: 10.1038/nri2779. 23

10.   Vishwakarma, A., Bhise, N. S., Evangelista, M. B., Rouwkema, J., Dokmeci, M. R.,
        Ghaemmaghami, A. M., Ghaemmaghami, A. M., Vrana, M. E., and Khademhosseini,
        A. (2016). Engineering immunomodulatory biomaterials to tune the inflammatory re-
        sponse, *Trends Biotechnol.* 34(6):470-82. DOI: 10.1016/j.tibtech.2016.03.009. 23, 24

11.   Davies, L. C., Jenkins, S. J., Allen, J. E., and Taylor, P. R. (2013). Tissue-resident macro-
        phages. *Nat. Immunol,* 14(10):986-95. DOI: 10.1038/ni.2705. 23

12.   Lee, J., Byun, H., Madhurakkat Perikamana, S. K., Lee, S., and Shin, H. (2019). Current
        advances in immunomodulatory biomaterials for bone regeneration, *Adv. Healthcare Mat.*
        8(4):1801106. DOI: 10.1002/adhm.201801106. 23, 24

13.   Chandorkar, Y., Ravujynarm K., and Basu, B. (2019). The Foreign Body Response Demy-
        stified, *ACS Biomat. Sci. Eng.* 5(1):19-44. DOI: 10.1021/acsbiomaterials.8b00252. 24, 28

14.   Mariani, E., Lisignoli, G., Borzi, R. M., and Pulsatelli, L. (2019). Biomaterials: For-
        eign bodies or tuners for the immune response?, *Int. J. Mol. Sci.* 20(3). DOI: 10.3390/
        ijms20030636. 24, 28

15.   Cravedi, P., Farouk, S., Angeletti, A., Edgar, L., Tamburrini, R., Duisit, J., Perub, L., and
        Orlando, G. (2017). Regenerative immunology: the immunological reaction to biomate-
        rials, *Trans. Intl.* 30(12):1199-208. DOI: 10.1111/tri.13068. 24

16.   Hume, P. S., He, J., Haskins, K., and Anseth, K. S. (2012). Strategies to reduce dendritic
        cell activation through functional biomaterial design, *Biomaterials* 33(14):3615-25. DOI:
        10.1016/j.biomaterials.2012.02.009. 24

17.   Spiller, K. L., Nassiri, S., Witherel, C. E., Anfang, R. R., Ng, J., Nakazawa, K. R., Yu, T.,
        and Vunjak-Novakovic, G. (2015). Sequential delivery of immunomodulatory cytokines
        to facilitate the M1-to-M2 transition of macrophages and enhance vascularization of
        bone scaffolds, *Biomaterials.* 37:194-207. DOI: 10.1016/j.biomaterials.2014.10.017. 24

18.   Chen, J., Li, M., Yang, C., Yin, X., Duan, K., Wang, J., and Feng, B. (2018). Macrophage
        phenotype switch by sequential action of immunomodulatory cytokines from hydrogel
        layers on titania nanotubes, *Colloids Surf. Biointerf.* 163:336-45. DOI: 10.1016/j.col-
        surfb.2018.01.007. 24

19.   Peng, Y., Tellier, L. E., and Temenoff, J. S. (2016). Heparin-based hydrogels with tunable
        sulfation & degradation for anti-inflammatory small molecule delivery, *Biomater. Sci.*
        4(9):1371-80. DOI: 10.1039/C6BM00455E. 24

20. Zhong, Y. and Bellamkonda, R. V. (2007). Dexamethasone-coated neural probes elicit attenuated inflammatory response and neuronal loss compared to uncoated neural probes, *Brain Res.* 1148:15-27. DOI: 10.1016/j.brainres.2007.02.024. 24

21. Udipi, K., Ornberg, R. L., Thurmond II, K. B., Settle, S. L., Forster, D., and Riley, D. (2000). Modification of inflammatory response to implanted biomedical materials *in vivo* by surface bound superoxide dismutase mimics, *J. Biomed.Mat. Res.* 51(4):549-60. DOI: 10.1002/1097-4636(20000915)51:4<549::AID-JBM2>3.0.CO;2-Z. 24

22. Swartzlander, M. D., Blakney, A. K., Amer, L. D., Hankenson, K. D., Kyriakides, T. R., and Bryant, S. J. (2015). Immunomodulation by mesenchymal stem cells combats the foreign body response to cell-laden synthetic hydrogels, *Biomaterials* 41:79-88. DOI: 10.1016/j.biomaterials.2014.11.020. 24

23. Chen, Z., Wu, C., Gu, W., Klein, T., Crawford, R., and Xiao, Y. (2014). Osteogenic differentiation of bone marrow MSCs by β-tricalcium phosphate stimulating macrophages via BMP2 signalling pathway, *Biomaterials* 35(5):1507-18. DOI: 10.1016/j.biomaterials.2013.11.014. 24

24. Bridges, A. W. and García, A. J. (2008). Anti-inflammatory polymeric coatings for implantable biomaterials and devices, *J. Diab. Sci. Tech.* 2(6):984-94. DOI: 10.1177/193229680800200628. 25

25. Bridges, A. W., Singh, N., Burns, K. L., Babensee, J. E., Andrew Lyon, L, and García, A. J. (2008). Reduced acute inflammatory responses to microgel conformal coatings, *Biomaterials.* 29(35):4605-15. DOI: 10.1016/j.biomaterials.2008.08.015. 25

26. Wong, T.-S., Kang, S. H., Tang, S. K. Y., Smythe, E. J., Hatton, B. D., Grinthal, A., and Aizenberg, J. (2011). Bioinspired self-repairing slippery surfaces with pressure-stable omniphobicity, *Nature* 477(7365):443-7. DOI: 10.1038/nature10447. 25

27. Kou, P. M., Pallassana, N., Bowden, R., Cunningham, B., Joy, A., Kohn, J., and Babensee, J. E. (2012). Predicting biomaterial property-dendritic cell phenotype relationships from the multivariate analysis of responses to polymethacrylates, *Biomaterials* 33(6):1699-713. DOI: 10.1016/j.biomaterials.2011.10.066. 25

28. Kajahn, J., Franz, S., Rueckert, E., Forstreuter, I., Hintze, V., Moeller, S., and Simon, J. C. (2012). Artificial extracellular matrices composed of collagen I and high sulfated hyaluronan modulate monocyte to macrophage differentiation under conditions of sterile inflammation, *Biomatter* 2(4):226-73. DOI: 10.4161/biom.22855. 25

29. Brown, B. N., Londono, R., Tottey, S., Zhang, L., Kukla, K. A., Wolf, M. T., Daly, K. A., Reing, J. E., and Badylak, S. F. (2012). Macrophage phenotype as a predictor of con-

structive remodeling following the implantation of biologically derived surgical mesh materials, *Acta biomaterialia* 8(3):978-87. DOI: 10.1016/j.actbio.2011.11.031. 25

30.   Lin, C.-C., Metters, A. T., and Anseth, K. S. (2009). Functional PEG–peptide hydrogels to modulate local inflammation inducedby the pro-inflammatory cytokine TNFα, *Biomaterials* 30(28):4907-14. DOI: 10.1016/j.biomaterials.2009.05.083. 26

31.   Vegas, A. J., Veiseh, O., Doloff, J. C., Ma, M., Tam, H. H., Bratlie, K., Bader, A. R., Langan, E., Olejik, K. Fenton, P. Kang, J. W., Hollister-Locke, J., Bochenek, M. A., Chiu, A., Siebert, S., Tang, K., Jhunjhunwala, S., Aresta-Dasilva, S., Dholakia, M., Thakrar, R., Vietti, T., Chen, M., Cohen, J., Siniakowicz, K., Qi, M., McGarrigle, J., Graham, A. C., Lyle, S., Harlan, D. M., Greiner, D. L., Overholzer, J., Weir, G. C., Langer, R., and Anderson, D. G. (2016). Combinatorial hydrogel library enables identification of materials that mitigate the foreign body response in primates, *Nat. Biotechnol.* 34(3):345-52. DOI: 10.1038/nbt.3462. 26, 27

32.   Vegas, A. J., Veiseh, O., Gürtler, M., Millman, J. R., Pagliuca, F. W., Bader, A. R., Doloff, J. C., Li, J, Chen, M., Olejnik, K., Tam, H. H., Jhunjhunwala, S., Langan, E., Aresta-Dasliva, S., Gandham, S., McGarrigle, J. J., Bochenek, M. A., Hollister-Lock, J., Oberholzer, J., Greiner, D. L., Weir, G. C., Melton, D. A., Langer, R., and Anderson, D. G. (2016). Long-term glycemic control using polymer-encapsulated human stem cell–derived beta cells in immune-competent mice, *Nat. Med.* 22(3):306-11. DOI: 10.1038/nm.4030. 26

33.   Bygd, H. C. and Bratlie, K. M. (2016). The effect of chemically modified alginates on macrophage phenotype and biomolecule transport., *J. Biomed. Mat. Res. Part A* 104(7):1707-19. DOI: 10.1002/jbm.a.35700. 26

34.   Jiang, S. and Cao, Z. (2010). Ultralow-fouling, functionalizable, and hydrolyzable zwitterionic materials and their derivatives for biological applications, *Adv. Mat.* 22(9):920-32. DOI: 10.1002/adma.200901407. 26

35.   Yesilyurt, V., Veiseh, O., Doloff, J. C., Li, J., Bose, S., Xie, X., Bader, A. R., Chen, M., Webber, M. J., Vegas, A. J., Langer, R., and Anderson, D. G. (2017). A facile and versatile method to endow biomaterial devices with zwitterionic surface coatings, *Adv. Healthcare Mat.* 6(4):1601091. DOI: 10.1002/adhm.201601091. 27

36.   Xie, X., Doloff, J. C., Yesilyurt, V., Sadraei, A., McGarrigle, J. J., Omami, M., Veiseh, O., Farah, S., Isa, D., Ghani, S., Joshi, I., Vegas, A., Li, J., Wang, W., Bader, A., Tam, H. H., Tao, J., Chen, H.-J., Yang, B., Williamson, K. A., Oberholzer, J., Langer, R., and Anderson, D. G. (2018). Reduction of measurement noise in a continuous glucose monitor by coating the sensor with a zwitterionic polymer., *Nat. Biomed. Eng.* 2(12):894-906. DOI: 10.1038/s41551-018-0273-3. 27

37.   Vassey, M. J., Figueredo, G. P., Scurr, D. J., Vasilevich, A. S., Vermeulen, S., Carlier, A., Luckett, J., Beijer, N. R. M., Williams, P., Winkler, D. A., de Boer, J., Ghaemmaghami, A. M., and Alexander, M. R. (2020). Immune modulation by design: using topography to control human monocyte attachment and macrophage differentiation, *Adv. Sci.* n/a(n/a):1903392. DOI: 10.1002/advs.201903392. 28, 29

38.   Roach, P., Eglin, D., Rohde, K., and Perry, C. C. (2007). Modern biomaterials: a review—bulk properties and implications of surface modifications, *J. Mat. Sci. Mat. in Med.* 18(7):1263-77. DOI: 10.1007/s10856-006-0064-3. 28

39.   Chen, S., Jones, J. A., Xu, Y., Low, H.-Y., Anderson, J. M., and Leong, K. W. (2010). Characterization of topographical effects on macrophage behavior in a foreign body response model, *Biomaterials* 31(13):3479-91. DOI: 10.1016/j.biomaterials.2010.01.074. 28

40.   Hulander, M., Lundgren, A., Berglin, M., Ohrlander, M., Lausmaa, J., and Elwing, H. (2011). Immune complement activation is attenuated by surface nanotopography, *Int. J. Nanomed* 6:2653-66. DOI: 10.2147/IJN.S24578. 28

41.   Jahed, Z., Molladavoodi, S., Seo, B. B., Gorbet, M., Tsui, T. Y., and Mofrad, M. R. K. (2014). Cell responses to metallic nanostructure arrays with complex geometries, *Biomaterials* 35(34):9363-71. DOI: 10.1016/j.biomaterials.2014.07.022. 28

42.   McWhorter, F.Y., Wang, T., Nguyen, P., Chung, T., and Liu, W. F. (2013). Modulation of macrophage phenotype by cell shape, *Proceedings of the National Academy of Sciences* 110(43):17253. DOI: 10.1073/pnas.1308887110. 28

43.   Park, J., Bauer, S., von der Mark, K., and Schmuki, P. (2007). Nanosize and vitality: TiO2 nanotube diameter directs cell fate, *Nano Lett.* 7(6):1686-91. DOI: 10.1021/nl070678d. 28, 29

44.   Alfarsi, M. A., Hamlet, S. M., and Ivanovski, S. (2014). Titanium surface hydrophilicity modulates the human macrophage inflammatory cytokine response, *J. Biomed. Mat. Res. Part A* 102(1):60-7. DOI: 10.1002/jbm.a.34666. 28

45.   Rostam, H. M., Singh, S., Vrana, N. E., Alexander, M. R., and Ghaemmaghami, A. M. (2015). Impact of surface chemistry and topography on the function of antigen presenting cells, *Biomat. Sci.* 3(3):424-41. DOI: 10.1039/C4BM00375F. 28

46.   Kusaka, T., Nakayama, M., Nakamura, K., Ishimiya, M., Furusawa, E., and Ogasawara, K. (2014). Effect of silica particle size on macrophage inflammatory responses, *PLOS ONE.* 9(3):e92634. DOI: 10.1371/journal.pone.0092634. 29

47.   Zandstra, J., Hiemstr, C., Petersen, A. H., Zuidema, J., van Beuge, M. M., Rodriguez, S,. Lathuile, A. A., Veldhuis, G. J., Steendam, R., Bank, R. A., and Popa, E. R. (2014). Microsphere size influences the foreign body reaction, *Eur. Cell Mater.* 28:335-47. DOI: 10.22203/eCM.v028a23. 29

48.   Ward, W. K., Slobodzian, E. P., Tiekotter, K. L., and Wood, M. D. (2002). The effect of microgeometry, implant thickness and polyurethane chemistry on the foreign body response to subcutaneous implants, *Biomaterials* 23(21):4185-92. DOI: 10.1016/S0142-9612(02)00160-6. 29

49.   Matlaga, B. F., Yasenchak, L. P., and Salthouse, T. N. (1976). Tissue response to implanted polymers: The significance of sample shape, *J. Biomed. Mat. Res.* 10(3):391-7. DOI: 10.1002/jbm.820100308. 29

50.   Veiseh, O., Doloff, J. C., Ma, M., Vegas, A. J., Tam, H. H., Bader A., R, Li, J., Langan, E., Wyckoff, J., Loo, W. S., Jhunjhunwala, S., Chiu, A., Siebert, S., Tang, K., Hollister-Lock, J., Aresta-Dasilva, S., Bochenek, M., Mendoza-Elias, J., Wang, U., Qi, M., Lavin, D. M., Chen, M., Dholakia, N., Thakrar, R., Lacik, I., Weir, G. C., Overholzer, J., Greiner, D. L., Langer, R., and Anderson, D. G. (2015). Size- and shape-dependent foreign body immune response to materials implanted in rodents and non-human primates, *Nat. Mat.* 14(6):643-51. DOI: 10.1038/nmat4290. 29, 31

51.   Rajyalakshmi, A., Ercan, B., Balasubramanian, K., and Webster, T. J. (2011). Reduced adhesion of macrophages on anodized titanium with select nanotube surface features, *Int. J. Nanomed.* 6:1765-71. DOI: 10.2147/IJN.S22763. 29

CHAPTER 4

Immunomodulatory Organic and Polymer Nanomedicine in Cancer Therapy

Vijay Sagar Madamsetty, Anubhab Mukherjee, Sudip Mukherjee

4.1 IMMUNOMODULATORY ORGANIC AND POLYMER NANOMEDICINE

In the past two decades, cancer immunotherapy has developed rapidly. Recent accomplishments in cancer immunotherapy have created momentous interest in harnessing the body's immune system to fight cancer [1]. Numerous methodologies have been investigated to improve the effectiveness while minimizing toxicities of cancer immunotherapy [2]. Various cancer immunotherapies, such as therapeutic cancer vaccines, adoptive cell therapy, or immune checkpoint blocking therapy, either used alone or combined with other therapies [3]. However, so far, the clinical outcome of most cancer immunotherapy is not adequate. It has been widely documented that nanotechnology can improve the efficacy of cancer immunotherapy [4]. A variety of nanoparticles have been developed, which have excellent immunogenicity and modifiability, and can carry tumor therapeutic drugs to achieve combined therapy, to improve the effectiveness and durability of antitumor immunity while reducing adverse side effects [4]. Traditionally used nanomaterials are also being explored to boost host anticancer immunity [5]. Different nanoformulations of antigens, chemokines, cytokines, oligonucleotides, and Toll-like receptor (TLR) agonists targeting many immune cells have been effectively established in many preclinical settings, producing promising results [6]. This chapter will focus on organic and polymeric nanoparticles-based immunomodulatory effects and immuno-stimulatory agents' transport to help horse antitumor immune responses.

Research in organic and polymer based-nanotechnology has resulted in noticeable progress in cancer treatment [7]. Numerous nanoparticles have been developed using organic materials, which are more interested in synthesizing and preparing these nanoparticles. Due to the base organic molecules' excellent physiochemical and biological properties, these nanoparticles have their own advantages in cancer therapy in boosting the immune system [8]. The advantageous properties of frequently used base materials and functional performance of the various organic NPs *in vivo* are discussed. Polymeric nanoparticles, such as micelles dendrimers and polymer-drug conjugates, have expected increasing attention for cancer treatment since they can effectively deliver anticancer agents to the tumor by passive or active targeting mechanisms [9]. Owing to their important unique characteristics, micelles containing polymeric amphiphiles can solubilize the hydrophobic drugs and manipulate their *in vivo* fates. The polymer backbone's functional groups facilitate the chemical conjugation of specific ligands, allowing for anticancer drugs' targeted delivery. These advantages of polymeric nanomaterials make them attractive to design formulations for targeted drug delivery to enhance therapeutic efficacy.

Meanwhile, in the early 2000s, much effort was devoted to cancer immunotherapies using polymeric nanomedicines that deliver antigen to DCs for the antigen-specific immune response or siRNA for immune cell reprogramming [10]. After the advanced success of immune checkpoint inhibitors in the clinic, there have been many reports on the combined effects of cancer immunotherapy and polymeric nanomedicine [11]. In recent years, polymeric nanomedicines have gained increasing recognition for cancer immunotherapy since they can enhance the immunotherapeutic efficacy and surmount limitations of Immune-Related Adverse Events (IRAEs) [12]. For example, polymeric nanomedicine allows for targeted delivery of cancer antigens to antigen-presenting cells (APCs), such as macrophages and dendritic cells (DCs), by which immature T cells can be effectively primed [13, 14]. The adverse events of immune checkpoint therapies, induced by their binding to nontarget cells, can be reduced by employing nanomedicine technology [15]. Also, polymeric nanomedicine is useful to elicit an appropriate immune response by delivering cytokines to effector T cells. In recent years, polymeric nanomedicines before cancer immunotherapy have been extensively investigated, as highlighted in this chapter.

4.2   SYNTHESIS AND CHARACTERIZATION TECHNIQUES

Nanotechnology is an emerging field of science. The base of nanotechnology is nanoparticles which range in size from 100–200 nm [16]. Organic nanoparticles are made up of lipids, proteins, carbohydrates, and other organic frameworks and are classified as graphene, fullerene, carbon nanotubes, carbon nanofiber, and carbon black nanoparticles based on carbon structure. They are also categorized as one-dimensional nanoparticles, two-dimensional nanoparticles, and three-dimensional nanoparticles based on dimensions. The organic nanoparticles, such as ferritin, micelles, dendrimers, and liposomes, are non-toxic; being biodegradable is ideal for drug delivery due to flexible and easy tunable characteristics [17]. Organic nanoparticles are expected to have more potential than other inorganic and metallic nanoparticles due to variability in the synthesis, and modification of organic compounds is infinite. Thus, small molecular organic nanostructures have fascinating potential as they can compromise easily tunable properties through molecular design. The organic and polymer nanoparticles are synthesized using different methods based on the purpose and materials used. These nanoparticles' formation depends on critical parameters such as molecular weight, charge, or stabilizer choice [18]. These parameters also control the nanoparticle's overall mean size and surface, a critical feature for drug delivery. For example, various procedures can be used to develop polymer nanoparticles, such as solvent evaporation, nanoprecipitation, salting-out, dialysis, supercritical fluid technology, micro-emulsion, mini-emulsion, surfactant-free emulsion, and interfacial polymerization.

Solvent Evaporation

The solvent evaporation method is a widely used method to prepare the polymeric nanoparticles by biodegradable polymers that have been applied in the drug delivery systems [19]. The used polymer and drug are dissolved in an organic solvent such as dichloromethane/chloroform/ethyl acetate, and emulsions are formulated using the aqueous solution with surfactants. The solvent evaporation method has two main approaches that are used for the formation of emulsions. The first one is the preparation of single emulsions [(oil-in-water (o/w)] and the second one is double emulsions [(water-in-oil-in-water, (w/o/w)]. In this method, high-speed homogenization or ultrasonication is used for the preparation of emulsion. After that, the organic solvent is evaporated, and the nanoparticles are collected by ultracentrifugation. The obtained nanoparticles are washed with distilled water for removing surfactants. Even though the solvent evaporation method is a simple and easy method for preparing polymeric nanoparticles, the time-consuming and possible agglomeration of the nanoparticles during the evaporation process may affect nanoparticles' particle size and morphology.

Nanoprecipitaion Method

Nanoprecipitation is the most used method for the synthesis of organic polymer nanoparticles. Nanoprecipitation requires the organic solvent to be miscible with water. In the nanoprecipitation method, an organic (e.g. THF, acetone) solution of the polymer is emulsified in an aqueous solution (with or without surfactant) using sonication or stirring [20]. The organic solvent is then removed by stirring (with or without vacuum), and in the process of decreasing solvent, hydrophobicity leads to aggregation of polymers and hydrophobic segments of the encapsulation matrix to yield nanoparticles. The polymeric matrix's hydrophilic chains orient into the aqueous phase to facilitate further functionalization of nanoparticles.

Salting Out

Salting out is a method for preparing polymeric nanoparticles established on separating a water-miscible solvent like acetone/THF/ethanol using a salting-out effect [21]. The salting-out method modifies the emulsification by beginning with an emulsion. The organic phase is prepared by dissolving the polymer in a solvent (acetone, tetrahydrofuran, and ethanol) and then added to the aqueous phase (water), the salting-out agent (calcium chloride, magnesium chloride, sucrose) and a stabilizer. Sufficient water volume is added to the mixture, and the resulting nanoparticles are collected via cross-flow filtration. The salting-out method has some advantages that are the most important of reducing stress to protein encapsulants. Another advantage is that it does not occur in an increase in temperature. Therefore, heat-sensitive molecules can be processed by this method. To the contrary of advantages, this method has some disadvantages that are specific applications to lipophilic drugs and extensive washing steps.

Coacervation

Coacervation involves mixing two oppositely charged polyelectrolytes and the addition of organic polymers in an alkaline aqueous solution [22]. Nanoparticles form due to the precipitation of organic polymers at an alkaline pH. The polymer-rich phase and electrolytes form a barrier that covers the particles forming a droplet. These droplets are purified by filtration and ultracentrifugation, followed by a neutralization step through washing to form the organic polymer nanoparticles.

Dialysis

Dialysis is a preparation method that provides a basic and effective way to prepare small polymer nanoparticles (PNPs) [23]. First, the polymer is dissolved in the organic solvent then put inside a dialysis tube while producing the nanoparticle by dialysis. It is important to make sure that the dialysis tube is suitable for the molecular weight of the nanoparticle. During dialysis, the solvent loses its solubility as a result of displacement. In this way, the polymer's progressive aggregation occurs, and a homogeneous suspension of nanoparticles is obtained. Likewise, radical polymerization is used to prepare nanoparticles with poly-alkyl (cyano) acrylate. The polymer is liquefied in an aqueous solution, followed by the cyanoacrylate's addition, which follows the cyanoacrylate dispersion polymerization for particle formation.

Similarly, lipid-based organic nanoparticles are synthesized by various methods, including the thin-film hydration method, reverse phase evaporation method, solvent injection method, detergent removal method, supercritical reverse phase evaporation method [24], supercritical anti-solvent method, and crossflow injection method. The details of these preparation methods were discussed in the published literature.

Characterization of Nanoparticles

Characterization of nanoparticles is based on analysis of particle size, polydispersity index (PDI), zeta potential, surface area, morphology, and composition [25]. Particle size and distribution are the main characteristics of nanoparticles, determining the significant drug loading, drug release, and nanoparticles' stability [26, 27]. Zeta potential is another important feature used to describe the surface charge property of the nanoparticles. Dynamic light scattering (DLS) is used for measuring the size, distribution, and zeta potentials of the developed nanoparticles. Recently, atomic force microscopy (AFM), transmission electron microscopy (TEM), and scanning electron microscopy (SEM) are also being used for the characterization of nanoparticles by looking at their morphological studies [28]. X-ray photoelectron spectroscopy (XPS) is used for surface chemistry analysis of nanoparticle suspension. Some other techniques like Fourier Transform Infrared Spectroscopy (FT-IR), UV-Vis spectrophotometry, X.ray diffraction (XRD), Annular Dark-Field Imaging (HAADF), and Intracranial Pressure (ICP) are using for measuring different application synthesized nanoparticles [29].

Dynamic Light Spectroscopy (DLS)

DLS is a powerful tool for studying nanoparticles size and diffusion behavior in solution [30]. DLS calculates the size of the nanoparticle in a solvent-based on the Brownian motion. The Stokes–Einstein equation is used to determine the hydrodynamic diameter of the diffusion factor in a homogeneous solution. DLS-based measurement is appropriate for monodisperse and polydisperse nanomaterials but not suitable for high-dimensional samples. The diffusion coefficient and hydrodynamic radii are calculated based on nanoparticles size and shape. The value of the zeta potential also offers evidence about the stability of the colloidal solution. The stability of colloidal particles varies with the type and rate of interaction between the particles. If all the suspension particles have a massive negative or positive zeta potential, they push each other and agglomeration or precipitation do not occur. When the zeta potential is low, there is not enough force to keep the particles away from each other, and agglomeration occurs. DLS helps to study the homogeneity of proteins, nucleic acids, and complexes of protein-protein or protein-nucleic acid preparations and study protein-small molecule interactions. The DLS technique's main advantage is that measurements can be made quickly and at a low cost.

Transmission Electron Microscopy (TEM)

Transmission electron microscopy (TEM) is a valuable technique for the physico-chemical characterization of newly synthesized nanoparticles [31]. It also helps to explore the effects of nanocomposites on biological systems, providing essential information for efficient therapeutic and diagnostic strategies. TEM uses an electron beam to image a nanoparticle sample, providing much higher resolution than is possible with light-based imaging techniques. TEM is the ideal method to measure nanoparticle size, distribution, and morphology. Thus, for the development of nanotechnology in the biomedical field, experts in cell biology, histochemistry, and ultramicroscopy should always support the chemists, physicists, and pharmacologists engaged in the synthesis and characterization of innovative nano constructs. Sample preparation for the TEM is complicated and time-consuming because it needs to be a very thin sample for electron affinity. Nanoparticle sample is precipitated in the films during TEM characterization, then immobilized with staining material. This process provides withstanding against vacuum. When an electron bar is transmitted through an ultra-fine, it contacts the sample, and the sample's surface characteristics are obtained.

Atomic Force Microscopy (AFM)

Atomic force microscopy (AFM) is a powerful, multifunctional imaging technique for surface analysis of micro/nanoparticles or any biological samples, from single molecules to living cells, to be visualized [32]. This flexible technique can be used to obtain high-resolution nanoscale images and study local sites in the air (conventional AFM) or liquid (electrochemical AFM) surroundings. The AFM creates an image based on the surface forces between the sample surface and the tip attached to an arm. The sharp tip used in this technique is 1–2 μm long and less than 100 Å in diameter. The forces that occur between the surface of the material and the inspected tip during a probe travel cause to diverge the tip's position. The surface topography is created by measuring this deviation. Atomic force microscope can be operated in three different ways. The first one is the *contact* type, where the sharp tip makes soft physical contact with the surface, and the changes in the arm's position are recorded. In the second *non-contact* type, the arm vibrates in a position close to the sample surface (50–150 Å). In this method, the attractive Van der Waals forces acting, and depending on the changes in the surface, the changes occurring in the frequency of the arm's vibration are measured. The third one, the *tapping* method, is similar to the contact method. This powerful technique allows high-resolution topographic imaging of sample surfaces. Similarly, the other methods are also used to characterize based on the application.

4.3   RECENT EXAMPLES OF IMMUNOMODULATORY ORGANIC AND POLYMER NANOMATERIALS IN CANCER THERAPY

Various kinds of organic-based nanomaterials, including liposome, micelles, dendrimers polymer-based, are being applied for cancer immunotherapy [33]. Along with recent advances in immunology, investigators began to explore nanotechnology's potential for improving cancer immunotherapy. In contrast to conventional nanomedicine, which targets cancer cells, cancer immunotherapy nanoparticles can also target dendritic cells or lymph nodes or APCs in circulating blood or lymphoid tissues. The activated immune cells can travel along with chemokine gradients to tumors and kill them by this method. It is well established that nanoparticle-based delivery of T cells or natural killer cells can work effectively in tumors with high concentrated drugs than nanoparticles alone. Additionally, it is well demonstrated that an excessively high cytotoxic drug concentration needed to be delivered into tumors to succeed complete tumor eradication, which is very difficult. In contrast, a much lower immunomodulators concentration is required for a robust anticancer immune response, as stimulated immune cells can proliferate and expand with a memory effect. The nanoparticles mediated cancer immunotherapy overcomes several limitations in other nanotherapies, such as avoid opsonization and reduce unwanted phagocytes uptake. The FDA approved PLGA is one of the most applied polymeric-based nanomaterials used in cancer therapy due to its multifunction, biodegradation, and protection of agents for delivery and applications for TME modulation. For example, Dölen et al. demonstrated the efficiencies PLGA nanoparticles decorated with TLR ligands and full-length ovalbumin (OVA) in combination with adjuvant α-galactosylceramide (α-GalCer) [34]. They found that the OVA + α-GalCer PLGA nanoparticles induced strong T cell-mediated antitumor immune responses, and the growth of established tumors delayed by OVA + α-GalCer nanoparticles. A similar study was recently carried out by Florindo and his team and demonstrated a robust tumor-suppressive effect in melanoma [35]. in their study, a combination of α-galactosylceramide and TLR ligands within PLGA nanoparticle-induced a prominent anti-tumor immune response able to restrict melanoma growth. Enhanced infiltration of NKT and NK. T cells into the tumor site were only achieved when the combination GalCer, antigens, and TLR ligands was co-delivered by the nanovaccine (35). Scientists developed PLGA-based core-shell nanoparticles loaded with both a water-soluble enzyme (catalase), which can decompose H_2O_2 to generate O_2 and hydrophobic imiquimod. These PLGA-R837@Cat nanoparticles greatly enhanced radiotherapy effectiveness by modulating the immune-suppressive tumor microenvironment and dismissing tumor hypoxia [36].

Chen et al. integrated TLR7 agonist (imiquimod) into PLGA nanoparticles and encapsulated photothermal agents (indocyanine green) for combination therapy [37]. In this study, the authors used polyethylene glycol-grafted PLGA co-polymer (mPEG-PLGA) was used to co-load both ICG and R837. These nanoparticles showed vaccine-like functions in *in vivo* settings and demonstrated enhanced anti-tumor activity in the C26 tumor model when used in combination with CTLA-4 checkpoint inhibitors (Figure 4.1). Furthermore, these nanoparticles also showed strong immune response and memory effect for various types of tumor models

Using dual immune-modulating therapies has recently gained great attention as a potent cancer treatment strategy. A combination of immune therapies such as anti-programmed cell death-1 (aPD1) and aOX40 show promising results recently; however, their codelivery is not explored. Scientists are working on developing an efficient nano-drug delivery system to deliver these dual immune agents simultaneously. For example, investigators delivered aPD1 and aOX40 using polymeric-based nanoparticles, which enhanced T-cell activation and showed a more significant therapeutic effect with boosting the immune system [38]. Several polymeric-based nanoparticles have been developed for cancer immunotherapy and are cited here for the interest scientific community [39]. In summary, polymeric nanoparticles provided an outstanding contribution to delivering immune modulator agents and simultaneously improving the anti-tumor efficacy.

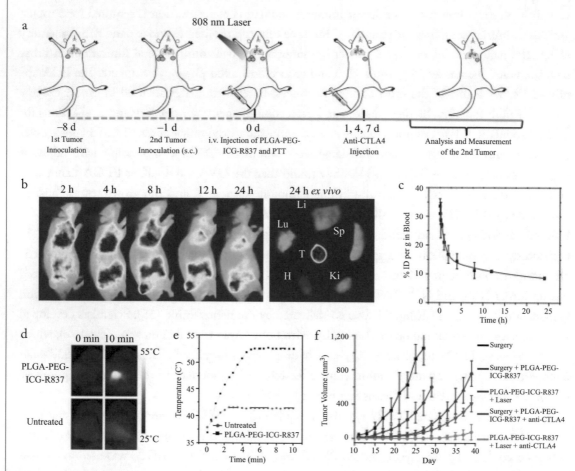

Figure 4.1: PTT-triggered immunotherapy via systemic injection of nanoparticles. (a) Schematic illustration showing the design of animal experiments. (b) *In vivo* fluorescence images of CT26-tumor-bearing mice taken at different time points post i.v. injection of PLGA-PEG-ICG-R837. The right column shows an ex vivo fluorescence image of major organs and tumor dissected from the mouse 24 h post injection. Tu, Li, Sp, Ki, H, and Lu stand for tumor, liver, spleen, kidney, heart, and lung, respectively. (c) Blood circulation curve of PLGA-PEG-ICG-R837 in mice by measuring the fluorescence of ICG in blood at different time points post i.v. injection (three mice per group). (d) IR thermal images of CT26-tumor-bearing mice injected with PLGA-PEG-ICG-R837 or PBS under the 808 nm laser (0.8 Wcm^2) irradiation. (e) The tumor temperature changes based on IR thermal imaging date in (d). (f) The growth curves of secondary tumors in different groups of CT26-tumor-bearing mice after various treatments to eliminate their primary tumors (six mice per group). Data are presented as the mean±s.e.m (reused with permission [37])

4.4   ADVANTAGES AND CHALLENGES

Like other nanoparticles, the great advantage of polymeric nanoparticles allows encapsulation of several bioactive small molecule drugs and shielding them against enzymatic or hydrolytic degradations. These nanoparticles form a thermodynamically stable colloidal solution with self-assembly of amphiphilic block copolymers. These copolymers in polymeric nanoparticles increase the encapsulated drug stability, controls the slow drug release, improves efficacy, and reduces systemic and chemotherapy-associated toxicity. Additionally, polymers have multiple advantages, such as modifying the nanoparticles' chemical, physical, and biological properties based on the requirement. These nanoparticles increase the pharmacokinetics of the encapsulated drugs with prolonged circulation time in the bloodstream. These are highly biodegradable that break down in physiological conditions, which minimize any potential toxicity of macromolecule and facilitate drug release. These nanoparticles also help in increasing the permeability and efficacy of drugs passing through biological barriers.

However, there are some disadvantages with these nanoparticles, like they tend to dissociate and premature drug release and interaction with proteins in circulation. These are too erudite to place on production, while the simple cannot meet the curative requirement. These nanoparticles also sometimes decrease cellular targeting capacity due to nonspecific serum protein binding and hinder tumor penetration. Hence, a pronounced need to develop efficient polymeric nanoparticles for better cancer immunotherapy translates to the clinic.

4.5   REFERENCES

1.   Dobosz, P. and Dzieciątkowski, T. (2019). The intriguing history of cancer immunotherapy, *Front. Immunol.* 10. DOI: 10.3389/fimmu.2019.02965. 37

2.   Riley, R. S., June, C. H., Langer, R., and Mitchell, M. J. (2019). Delivery technologies for cancer immunotherapy, *Nat. Rev. Drug Disc.* 18:175–96. DOI: 10.1038/s41573-018-0006-z. 37

3.   Mougel, A., Terme, M., and Tanchot, C. (2019). Therapeutic cancer vaccine and combinations with antiangiogenic therapies and immune checkpoint blockade, *Front. Immunol.* 10. DOI: 10.3389/fimmu.2019.00467. 37

4.   Yan, S., Zhao, P., Yu, T., and Gu, N. (2019). Current applications and future prospects of nanotechnology in cancer immunotherapy, *Cancer Biol. Med.* 16:486–97. DOI: 10.20892/j.issn.2095-3941.2018.0493. 37

5.   Liu, Z., Jiang, W., Nam, J., Moon, J. J., and Kim, B. Y. S. (2018). Immunomodulating nanomedicine for cancer therapy, *Nano Lett.* 18:6655–9. DOI: 10.1021/acs.nanolett.8b02340. 37

6. Piktel, E., Niemirowicz, K., Wątek, M., Wollny, T., Deptuł,a P., and Bucki, R. (2016). Recent insights in nanotechnology-based drugs and formulations designed for effective anti-cancer therapy, *J. Nanobiotechnol.* 14. DOI: 10.1186/s12951-016-0193-x. 37

7. Navya, P. N., Kaphle, A., Srinivas, S. P., Bhargava, S. K., Rotello, V. M., and Daima, H. K. (2019). Current trends and challenges in cancer management and therapy using designer nanomaterials, *Nano Converg.* 6:23. DOI: 10.1186/s40580-019-0193-2. 37

8. Luk, B. T., and Zhang, L. (2014). Current advances in polymer-based nanotheranostics for cancer treatment and diagnosis, *ACS Appl. Mat. Interf.* 6:21859–73. DOI: 10.1021/am5036225. 37

9. Senapati, S., Mahanta, A. K., Kumar, S., and Maiti, P. (2018). Controlled drug delivery vehicles for cancer treatment and their performance, *Signal. Transduct. Target. Ther.* 3:7. DOI: 10.1038/s41392-017-0004-3. 38

10. Leach, D. G., Young, S., and Hartgerink, J. D. (2019). Advances in immunotherapy delivery from implantable and injectable biomaterials, *Acta Biomaterialia* 88:15–31. DOI: 10.1016/j.actbio.2019.02.016. 38

11. Fan, Y., Zhang, C., Jin, S., Gao, Z., Cao, J., Wang, A., Li, D., Wang, Q., Sun, X., and Bai, D. (2019). Progress of immune checkpoint therapy in the clinic (Review), *Oncol. Rep.* 41:3–14. DOI: 10.3892/or.2018.6819. 38

12. Darnell, E. P., Mooradian, M. J., Baruch, E. N., Yilmaz, M., and Reynolds, K. L. (2020). Immune-related adverse events (irAEs): Diagnosis, management, and clinical pearls, *Curr. Oncol. Rep.* 22:39. DOI: 10.1007/s11912-020-0897-9. 38

13. Zhuang, J., Holay, M., Park, J. H., Fang, R. H., Zhang, J., and Zhang, L. (2019). Nanoparticle delivery of immunostimulatory agents for cancer immunotherapy, *Theranostics* 9:7826–48. DOI: 10.7150/thno.37216. 38

14. Xu, X., Li, T., Shen, S., Wang, J., Abdou, P., Gu, Z., and Mo, R. (2019). Advances in engineering cells for cancer immunotherapy, *Theranostics* 9:7889–905. DOI: 10.7150/thno.38583. 38

15. Kim, K. and Khang, D. (2020). Past, present, and future of anticancer nanomedicine, *Int. J. Nanomed.* 15:5719–43. DOI: 10.2147/IJN.S254774. 38

16. Jeevanandam, J., Barhoum, A., Chan, Y. S., Dufresne, A., and Danquah, M. K. (2018). Review on nanoparticles and nanostructured materials: history, sources, toxicity and regulations, *Beilstein J. Nanotechnol.* 9:1050–74. DOI: 10.3762/bjnano.9.98. 38

17. Chowdhury, A., Kunjiappan, S., Panneerselvam, T., Somasundaram, B., and Bhattacharjee, C. (2017). Nanotechnology and nanocarrier-based approaches on treatment of degenerative diseases, *Int. Nano Lett.* 7:91–122. DOI: 10.1007/s40089-017-0208-0. 38

18. Khan, I., Saeed, K., and Khan, I. (2019). Nanoparticles: Properties, applications and toxicities, *Arabian J. Chem.* 12:908–31. DOI: 10.1016/j.arabjc.2017.05.011. 39

19. Krishnamoorthy, K. and Mahalingam, M. (2015). Selection of a suitable method for the preparation of polymeric nanoparticles: multi-criteria decision making approach, *Adv. Pharm. Bull.* 5:57–67. DOI: 10.4103/2231-4040.137410. 39

20. Hornig, S., Heinze, T., Becer, C. R., and Schubert, U. S. (2009). Synthetic polymeric nanoparticles by nanoprecipitation, *J. Mat. Chem.* 19:3838–40. DOI: 10.1039/b906556n. 39

21. Zielińska, A., Carreiró, F., Oliveira, A. M., Neves, A., Pires, B., Venkatesh, D. N., Durazzo, A., Lucarini, M., Eder, P., Silva, A. M., Santini, A., and Souto, E. B. (2020). Polymeric nanoparticles: Production, characterization, toxicology and ecotoxicology, *Molecules* 25:3731. DOI: 10.3390/molecules25163731. 39

22. Ahlin Grabnar, P. and Kristl, J. (2011). The manufacturing techniques of drug-loaded polymeric nanoparticles from preformed polymers, *J. Microencapsul.* 28:323–35. DOI: 10.3109/02652048.2011.569763. 40

23. Nah, J.-W., Paek, Y.-W., Jeong, Y.-I., Kim, D.-W., Cho, C.-S., Kim, S.-H., and Kim, M.-Y. (1998). Clonazepam release from poly (DL-lactide-co-glycolide) nanoparticles prepared by dialysis method, *Arch. Pharm. Res.* 21:418–22. DOI: 10.1007/BF02974636. 40

24. Meure, L. A., Foster, N. R., and Dehghani, F. (2008). Conventional and dense gas techniques for the production of liposomes: a review, *AAPS Pharm. Sci. Tech.* 9:798–809. DOI: 10.1208/s12249-008-9097-x. 40

25. Danaei, M., Dehghankhold, M., Ataei, S., Hasanzadeh Davarani, F., Javanmard, R., Dokhani, A., Khorasani, S., and Mozafari, M. R. (2018). Impact of particle size and polydispersity index on the clinical applications of lipidic nanocarrier systems, *Pharmaceutics* 10:57. DOI: 10.3390/pharmaceutics10020057. 40

26. Masarudin, M. J., Cutts, S. M., Evison, B. J., Phillips, D. R., and Pigram, P. J. (2015). Factors determining the stability, size distribution, and cellular accumulation of small, monodisperse chitosan nanoparticles as candidate vectors for anticancer drug delivery: application to the passive encapsulation of [(14)C]-doxorubicin, *Nanotechnol. Sci. Appl.* 8:67–80. DOI: 10.2147/NSA.S91785. 41

27.   Caputo, F., Clogston, J., Calzolai, L., Rösslein, M., and Prina-Mello, A. (2019). Measuring particle size distribution of nanoparticle enabled medicinal products, the joint view of EUNCL and NCI-NCL. A step by step approach combining orthogonal measurements with increasing complexity, *J. Contr. Rel.* 299:31–43. DOI: 10.1016/j.jconrel.2019.02.030. 41

28.   Rydz, J., Šišková, A., and Andicsová Eckstein, A. (2019). Scanning electron microscopy and atomic force microscopy: Topographic and dynamical surface studies of blends, composites, and hybrid functional materials for sustainable future, *Adv. Mat. Sci. Eng.* 6871785. DOI: 10.1155/2019/6871785. 41

29.   Mourdikoudis, S., Pallares, R. M., and Thanh, N. T. K. (2018). Characterization techniques for nanoparticles: comparison and complementarity upon studying nanoparticle properties, *Nanoscale* 10:12871–934. DOI: 10.1039/C8NR02278J. 41

30.   Stetefeld, J., McKenna, S. A., and Patel, T. R. (2016). Dynamic light scattering: a practical guide and applications in biomedical sciences, *Biophys. Rev.* 8:409–27. DOI: 10.1007/s12551-016-0218-6. 41

31.   Malatesta, M. (2016). Transmission electron microscopy for nanomedicine: novel applications for long-established techniques, *Eur. J. Histochem.* 60:2751. DOI: 10.4081/ejh.2016.2751. 41

32.   Dufrêne, Y. F., Ando, T., Garcia, R., Alsteens, D., Martinez-Martin, D., Engel, A., Gerber, C., and Müller, D. J. (2017). Imaging modes of atomic force microscopy for application in molecular and cell biology, *Nat. Nanotechnol.* 12:295–307. DOI: 10.1038/nnano.2017.45. 42

33.   Mu, W., Chu, Q., Liu, Y., and Zhang, N. (2020). A review on nano-based drug delivery system for cancer chemoimmunotherapy., *Nano-Micro Letters* 12:142. DOI: 10.1007/s40820-020-00482-6. 42

34.   Dölen, Y., Kreutz, M., Gileadi, U., Tel, J., Vasaturo, A., van Dinther, E. A., van Hout-Kuijer, M. A., Cerundolo, V., and Figdor, C. G. (2016). Co-delivery of PLGA encapsulated invariant NKT cell agonist with antigenic protein induce strong T cell-mediated antitumor immune responses, *Oncoimmunology* 5:e1068493. DOI: 10.1080/2162402X.2015.1068493. 43

35.   Sainz, V., Moura, L. I. F., Peres, C., Matos, A. I., Viana, A. S., Wagner, A. M., Vela Ramirez, J. E., Barata, T. S., Gaspar, M., Brocchini, S., Zloh, M., Peppas, N. A., Satchi-Fainaro, R., and Florindo, H. F. (2018). α-Galactosylceramide and peptide-based

nano-vaccine synergistically induced a strong tumor suppressive effect in melanoma, *Acta Biomater.* 76:193–207. DOI: 10.1016/j.actbio.2018.06.029. 43

36.   Chen, Q., Chen, J., Yang, Z., Xu, J., Xu, L., Liang, C., Han, X., and Liu, Z. (2019). Nanoparticle-enhanced radiotherapy to trigger robust cancer immunotherapy, *Adv. Mat.* 31. DOI: 10.1002/adma.201802228. 43

37.   Chen, Q., Xu, L., Liang, C., Wang, C., Peng, R., and Liu, Z. (2016). Photothermal therapy with immune-adjuvant nanoparticles together with checkpoint blockade for effective cancer immunotherapy, *Nat. Commun.* 7:13193. DOI: 10.1038/ncomms13193. 43, 44

38.   Mi, Y., Smith, C. C., Yang, F., Qi, Y., Roche, K. C., Serody, J. S., Vincent, B. G., and Wang, A. Z. (2018). A dual immunotherapy nanoparticle improves t-cell activation and cancer immunotherapy, *Adv. Mat.* 30. DOI: 10.1002/adma.201706098. 43

39.   Lee, E. S., Shin, J. M., Son, S., Ko, H., Um, W., Song, S. H., Lee, J. A., and Park, J. H. (2019). Recent advances in polymeric nanomedicines for cancer immunotherapy, *Adv. Healthcare Mat.* 8:1801320. DOI: 10.1002/adhm.201801320. 43

CHAPTER 5

Immunomodulatory Liposomes in Cancer Therapy

Anubhab Mukherjee, Vijay Sagar Madamsetty, Sudip Mukherjee

5.1 IMMUNOMODULATORY LIPOSOMES

Ever since its discovery in 1965, liposomes have drawn significant attention in nanomedicine research due to their unique properties such as specific targeting, protection to their payload, high biocompatibility, and low toxicity. Liposomes are lipid vesicles containing a hydrophilic aqueous interior and lipophilic bilayer exterior [1]. The kinetics and dynamics of arrangement of phospholipids in water, a bilayer structure, is mainly driven by the hydrophobic effect which spatially organizes the amphiphiles so as to minimize entropically unfavorable interactions between hydrophobic carbon chains and the surrounding aqueous phase. The vesicular arrangement with a cocooned aqueous phase, in turn, render liposomes as carriers of both lipophilic and hydrophilic molecules. In other words, water soluble molecules can be encapsulated in the aqueous compartment and lipophilic molecules can be encapsulated into the lipid bilayer region. They have extensive biomedical applications. A plethora of biologically active compounds, including anticancer agents, antimicrobial agents, chelating agents, peptides, hormones, enzymes, proteins, vaccines, and genetic materials, have been evaluated for delivery using liposomes [2]. Almost 17 different liposomal formulations are clinically approved for use against many pathological indications and lot many are undergoing various phases of clinical trials. The clinical success of the liposomal systems can be attributed to their small nanometric particle size, high drug loading, low bilayer permeability, colloidal stability (as indicated by surface charge), surface modification by PEGylation for obtaining better pharmacological profile, and targetability [3, 4]. The majority of these products are used to improve drug delivery and reduce off-target toxicity associated with the incorporated cytotoxic drug [5, 6]. In addition to the delivery of traditional small molecules, liposomes have been used to deliver nucleic acids as well. For example, small interfering RNA-based drug Onpattro (Alnylam) got recent approval by the FDA and EMA [7].

It is evident from our discussion in Chapters 1 and 2 that tumor immunity, both humoral and cellular, can make use of a variety of immune cells. They can be classified into three major categories: messengers, killers, and regulators. Professional antigen-presenting cells—namely, dendritic cells (DC), macrophages, neutrophils, etc., are messengers [8]. Obviously, cytotoxic T lymphocytes and natural killer (NK) cells are known as killers which are responsible for destroying tumor cells [9, 10]. The third category is regulators, including regulatory T cells (T_{regs}), which are involved in inhibiting the activity of APCs and the proliferation and differentiation of effector T cells [11]. As many of the agents which can evoke an immune response are polypeptide, antibody, or nucleic acid drugs, liposomes are widely explored as an ideal carrier for them to mediate effective cancer immunotherapy by triggering humoral or cellular immune response [12]. This chapter provides a brief overview of the recent developments in liposomal NPs used for immunotherapy.

5.2 SYNTHESIS AND CHARACTERIZATION TECHNIQUES

Liposomal nanostructures can classically be synthesized by using thin film hydration method with varying mole ratios of phospholipids (such as HSPC, Soy-PC, Egg PC, DSPC, DMPC, DPPC, DOPC, POPC, SOPC, MSPC, etc.), PEGylated-phospholipids (such as DSPE-PEG(2000), etc.), cholesterol, and different drugs of choice, both hydrophobic and hydrophilic (such as Paclitaxel, oxaliplatin, etc.). In case of hydrophobic drugs, lipids and drugs are dissolved together in organic solvent (chloroform, dichloromethane, methanol, and ethanol when necessary, etc.). All ingredients are mixed homogeneously in a round-bottomed flask and an organic solvent gets evaporated by rotary evaporator resulting in a thin film. The lipid film is then kept under high vacuum for 3–4 hr to ensure complete removal of the organic content. Next, it is hydrated (by adding 1 mL DI water or any buffer of choice to it) for 1.0 h in a water bath just above the phase transition temperature of the lipids used. For hydrophilic drugs, the lipid layer is formed without the drug and it gets hydrated by the buffer containing the drug. PEGylation can be used for a better pharmacokinetic profile of the delivery system. Some ligands targeted to specific receptors in the cells can be covalently (or by other techniques) grafted onto the PEG chain. Unencapsulated drugs can be separated by either spontaneous precipitation or centrifugation or Amicon wash or gel-filtration technique. Liposomes are, in general, downsized by sequential extrusion through membrane pores with different sizes like 400 nm, 200 nm, 100 nm, 80 nm, etc. Bath and probe sonication can also be employed for size reduction of liposomes, i.e., production of unilamellar vesicles from multilamellar ones. Drug loading and % encapsulation efficiency of the drug of choice within the liposomal nanoparticles are determined by UV spectrophotometry or HPLC analysis. Average size, PDI and surface potential of the nanoparticles (NPs) are measured by DLS. Size and shape can also be visualized by image analysis—HR-TEM, Cryo-TEM, etc. Liposomes are generally lyophilized (5% lactose solution, etc. can be used as cryo-protectant) over 16–20 hrs. The amorphous white solid powder formed thereafter is reconstituted by adding the required volume of water. DLS study of this reconstituted formulation should reveal similar size, PDI, and surface potential of nanoparticles as it was before lyophilization [2, 13–16].

5.3   RECENT EXAMPLES OF IMMUNOMODULATORY LIPOSOMES IN CANCER THERAPY

There is no denying the fact that a multitude of attractive attempts were made in the recent past with liposomes to deliver immunotherapeutics—not only directly to tumors for immunogenic cell death (ICD), but also collectively to the infiltrating immune cells. We shall briefly discuss few categories.

5.3.1   DC-TARGETED LIPOSOMES

For a professional and proficient antigen presentation, DC needs to mature by recognizing pathogens and antigens via pathogen recognition receptors (PRRs), such as TLRs. In 2018, Guan et al. designed, synthesized, and characterized a novel immunostimulatory spherical nucleic acids (IS-LSNAs) of single-stranded RNA selective for toll-like receptors 7/8 (TLR 7/8 agonists). These nanostructures comprise of liposomal cores (with DOPC, etc.) functionalized with a dense shell of single-stranded RNA (ssRNA) tethered to the lipid bilayer via cholesterol. In subsequent experiments, these liposomal NPs activated TLR 7/8 via NF-kB signaling pathway after they were preferentially taken up by plasmacytoid DCs. Importantly, when the aqueous interior was loaded with ovalbumin peptide antigen, ova-specific $CD8^+$ T cells were primed [17] (Figure 5.1). Here, it is worth mentioning that TLR3 can also be explored to induce DC maturation [18]. With growing knowledge of the DC-based cancer immunotherapy, it has become so obvious to recognize that one of the most popular technique is to deliver TAAs to DCs to enhance antitumor immune responses. Nevertheless, delivery of DNA or RNA encoding TAA is also a very attractive and available alternative. In 2015, Markov et al. developed a mannose receptor-targeted nanocarriers to deliver EGFP-encoded plasmid DNA and whole tumor RNA to DCs. Using a murine melanoma model, the authors demonstrated that these liposomes yielded a five- to six-fold reduction in melanoma lung metastases and enhanced the antitumor response [19]. A similar approach was adopted by Garu et al. in 2016, where DC-targeting liposomes which carried melanoma antigen-encoded DNA (p-CMV-MART1) and as a result, protracted antimelanoma immune response was produced [20]. Surprisingly, while their lofty status was always maintained as a carrier, liposomes themselves are also demonstrated to be capable of inducing humoral and cellular immune responses. For example, archaeosomes (lipids from archaea) showed promise to activate DCs and provide adjuvant stimulation to the immune response [21].

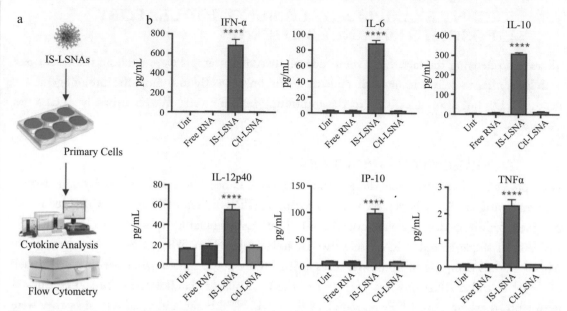

Figure 5.1: Fabrication of TLR7/8 agonist ssRNA-conjugated liposomal nanoparticles for DC-targeted cancer immunotherapy. Cytokine secretion by primary immune cells after 24 h of incubation with IS-LSNA NPs. Adapted with permission from [17]. ****P < 0.0001. Copyright (2018) John Wiley & Sons, Inc.

5.3.2   LIPOSOMES FOR T-CELLS

Cytotoxic T Lymphocytes expressing the glycoprotein CD8 on their cell surface are known as CD8$^+$+ T cells, their function being recognizing and terminating tumor cells. In addition, they also exude cytokines, like IFN-γ, to trigger tumor infiltration. Similarly, T cells expressing the glycoprotein CD4 on their surface are coined as CD4$^+$ T cells. They are called helper T cells as their key function is to assist other T cells in immunological processes, such as the activation of CTLs. In 2014, Korsholm et al. developed a formulation with cationic adjuvant and different antigens which induced antigen-specific CD8$^+$ T cells along with induction of effector-memory CD4$^+$ T cells [22]. It is also evident that agonists of a TNF receptor family can activate a strong cancer cell-killing response. In this regard, CD137, is a T-cell costimulatory molecule, expressed by activated T cells, which have gained research attention [23]. As mentioned earlier, Tregs inhibit the proliferation of CTLs by expressing CTLA-4 which, after binding with CD80 or CD86, shuts off the APC activity. For instance, an enhanced tumor accumulation and efficacy was observed for a PEGylated liposome with an encapsulated anti-CTLA-4 antibody [24]. It also has found its application in radio-immunotherapy [25].

5.3.3   LIPOSOMES FOR NK CELLS

NK cells represent humoral immunity as a subset of cytotoxic lymphocytes which can recognize cells under stress even without MHCs or antibodies resulting in a faster immune response. As of now, a number of reports have conveyed that overexpressed TRAIL on the surface of NK cell interacts with death receptors expressed on tumor cells and induce tumor cell apoptosis. For example, in order to avert lymphatic metastatic spread of subcutaneous tumor in mice, Chandrasekaran et al. developed liposomes decorated with TRAIL ligand and targeted NK cells by an anti-NK antibody. They found satisfactory results, needless to mention [26]. Exploration of CARs as a drug delivery platform is a powerful method to get rid of off-target effects. In 2017, Siegler and colleagues produced a modified CAR-NK cells cross-linking with multilamellar PTX-liposome by thiol-maleimide conjugation technique. This led to a precise tumor-targeting rendering it as a potent chemoimmunotherapy strategy [27]. More recently, DOX-encapsulated NK-92 cell fusogenic liposomes are reported to recognize cancer cels in an easier way and destroyed them [28].

5.3.4   LIPOSOMES FOR MACROPHAGES

We already had elaborately discussed about TAMs and their role in inflammation induced by the tumor in Chapter 2. A large number of studies have demonstrated that infiltrating macrophages mediate pancreatic cancer cell proliferation in invasive tumors. Few studies have confirmed that TAM depletion by liposomal clodronate can significantly reduce tumor metastasis [29]. Reprogramming M2 macrophage into M1 macrophages has been proposed for effective cancer therapy for their different behavior in tumor microenvironment. In 2019, Anderson et al. re-educated TAMs to turn into M1 phenotype from M2 phenotype by inhibiting the expression of STAT3-regulated genes. They developed corosolic acid (STAT3-inhibitory drug) encapsulated long-circulating liposomes to target CD163 receptor by decorating with an anti-CD163 antibody [30].

5.3.5   ADVANTAGES AND CHALLENGES

It is now comprehensible from our above discussion that the choice of biomaterials for delivery of therapeutics is very crucial, as this can directly affect the timeline for clinical translation. Stated differently, development of drug delivery systems using FDA-approved biomaterials are likely to enter the clinic and get approval faster than complex, unapproved ones. Complying with this, lipid-based systems are particularly useful for drug delivery as many are already approved by the FDA. Interestingly, one application using FDA-approved lipids to deliver mRNA to DCs, a tetravalent RNA-lipoplex cancer vaccine targeting four TAAs is now being evaluated in clinical trials for use in patients with advanced melanoma (NCT02410733) [31]. As mentioned earlier, Alnylam Pharmaceuticals is rejoicing FDA approval for their lipid–siRNA complex and setting a precedence for posterity to develop other RNA delivery platforms for cancer therapy, albeit no mRNA-based drug is still approved by the FDA [7, 32].

One major challenge before translatability of liposomal immunotherapeutics remains the choice of preclinical animal models employed for evaluating their potential. Few other critical criteria for clinical translation include stability of the vehicle, scalability, production cost, etc. [33]. Another daunting challenge remains targeting solid tumors for its immunosuppressive microenvironment, high interstitial fluid pressure, compressed vasculature, and dense fibrotic tissue in surrounding—together all inhibit T cell infiltration. Therefore, targeting the immune system and the tumor microenvironment could lead to more efficacious liposomal immunotherapies in the future.

5.4   REFERENCES

1.   Mukherjee, A., Waters, A. K., Kalyan, P., Achrol, A.S., Kesari, S., and Yenugonda, V.M. (2019). Lipid-polymer hybrid nanoparticles as a next generation drug delivery platform: State of the art, emerging technologies, and perspectives, *Int. J. Nanomed.* 14:1937–1952. DOI: 10.2147/IJN.S198353. 51

2.   Pattni, B. S., Chupin, V. V., and Torchilin, V. P. (2015). New developments in liposomal drug delivery, *Chem Rev.* 115 10938–10966. DOI: 10.1021/acs.chemrev.5b00046. 51, 52

3.   Sercombe, L., Veerati, T., Moheimani, F., Wu, S. Y., Sood, A. K., and Hua, S. (2015). Advances and challenges of liposome assisted drug delivery, *Front. Pharmacol.* 6:286. DOI: 10.3389/fphar.2015.00286. 51

4.   Bulbake, U., Doppalapudi, S., Kommineni, N. and Khan, W. (2017). Liposomal formulations in clinical use: an updated review, *Pharmaceutics* 9:1–33 DOI: 10.3390/pharmaceutics9020012. 51

5.   Din, F. U., Aman, W., Ullah, I., Qureshi, O. S., Mustapha, O., Shafique, S., and Zeb, A. (2017). Effective use of nanocarriers as drug delivery systems for the treatment of selected tumors, *Int. J. Nanomed.* 12:7291–7309. DOI: 10.2147/IJN.S146315. 51

6.   Patra, J. K., Das, G., Fraceto, L. F., Campos, E. V. R., Rodriguez-Torres, M. D. P., Acosta-Torres, L. S., Diaz-Torres, L. A., Grillo, R., Swamy, M. K., Sharma, S., Habtemariam, S., and Shin, H. S. (2018).Nano based drug delivery systems: recent developments and future prospects, *J. Nanobiotech.* 16(1):71. DOI: 10.1186/s12951-018-0392-8. 51

7.   Akinc, A., Maier, M. A., Manoharan, M., Fitzgerald, K., Jayaraman, M., Barros, S., Ansell, S., Du, X., Hope, M. J., Madden, T. D., Mui, B. L., Semple, S. C., Tam, Y. K., Ciufolini, M., Witzigmann, D., Kulkarni, J. A., van der Meel, R., and Cullis, P. R. (2019). The Onpattro story and the clinical translation of nanomedicines containing nucleic acid-based drugs, *Nat. Nanotechnol.* 14:1084–1087. DOI: 10.1038/s41565-019-0591-y. 51, 55

8.   den Haan, J. M., Aren, R., and van Zelm, M. C. (2014). The activation of the adaptive immune system: cross-talk between antigen-presenting cells, T cells and B cells, *Immunol Lett.* 162:103–12. DOI: 10.1016/j.imlet.2014.10.011. 52

9.   Beck, R. J., Slagter, M., and Beltman, J. B. (2019). Contact-dependent killing by cytotoxic T lymphocytes is insufficient for EL4 tumor regression *in vivo.*, *Cancer Res.* 79:3406–16. DOI: 10.1158/0008-5472. 52

10.   O'Leary, J. G., Goodarzi, M., Drayton, D. L., and von Andrian, U. H. (2006). T cell- and B cell independent adaptive immunity mediated by natural killer cells., *Nat Immunol.* 7:507. DOI: 10.1038/ni1332. 52

11.   Curie, T. J. (2007). Tregs and rethinking cancer immunotherapy, *J. Clin. Invest.* 117:1167–74. DOI: 10.1172/JCI31202. 52

12.   Blume, G., Cevc, G., Crommelin, M. D., Bakker-Woudenberg, I. A., Kluft, C., and Storm, G. (1993). Specific targeting with poly(ethylene glycol)-modified liposomes: coupling of homing devices to the ends of the polymeric chains combines effective target binding with long circulation times, *Biochim. Biophys. Acta.* 1149:180–4. DOI: 10.1016/0005-2736(93)90039-3. 52

13.   Roy, M., Biswas, G., Suryavanshi, H., Mukherjee, A., Kulkarni. A., and Sengupta, S. (2017). Cellular signaling inhibitors, their formulations and methods thereof, Publication no. WO2017/137958 PCT/IB2017/050770 (11.02.2017). 52

14.   Roy, M., Hossain, S., Sengupta, A., Mylavarapu, S., Sengupta, S., and Mukherjee, A. (2017). Supramolecular combinatorial therapeutics, Publication no. WO2015/153345 US2017/0112800 A1 (27.04.2017). 52

15.   Mukherjee, A., Bhattacharyya, J., and Chaudhuri, A. (2013). A Liposomal composition useful for inhibiting tumor growth through RNA-interference using liposomally associated CDC20 siRNA, Publication No: WO 2014/115158 Al. Patent Application No. 0202NF2012 PCT/IN2013/000054 (28/01/2013). 52

16.   Mukherjee, A., Bhattacharyya, J., Sagar, M. V., and Chaudhuri, A. (2013b) Liposomally encapsulated CDC20 siRNA inhibits both solid melanoma tumor growth and spontaneous growth of intravenously injected melanoma cells on mouse lung, *Drug Deliv. Transl. Res.* 3:224–234. DOI: 10.1007/s13346-013-0141-3. 52

17.   Guan, C. X., Chernyak, N., Dominguez, D., Cole, L., Zhang, B., and Mirkin, C. (2018). A. RNA-based immunostimulatory liposomal spherical nucleic acids as potent TLR7/8 modulators, *Small* 14:e1803284. DOI: 10.1002/smll.201803284. 53, 54

18.   Varypataki, E. M., Silva, A. L., Barnier-Quer, C., Collin, N., Ossendorp, F., and Jiskoot, W. (2016). Synthetic long peptide-based vaccine formulations for induction of cell mediated immunity: a comparative study of cationic liposomes and PLGA nanoparticles, *J. Control Release* 226:98–106. DOI: 10.1016/j.jconrel.2016.02.018. 53

19.   Markov, O. V., Mironova, N. L., Shmendel, E. V., Serikov, R. N., Morozova, N. G., Maslov, M. A., Vlassov, V. V., and Zenkova, M. A. (2015). Multicomponent mannose-containing liposomes efficiently deliver RNA in murine immature dendritic cells and provide productive anti-tumor response in murine melanoma model, *J. Control Release* 213:45–56. DOI: 10.1016/j.jconrel.2015.06.028. 53

20.   Garu, A., Moku, G., Gulla, S. K., and Chaudhuri, A. (2016). Genetic Immunization With *In vivo* Dendritic cell-targeting liposomal dna vaccine carrier induces long-lasting antitumor immune response, *Mol. Ther.* 24(2):385-397. DOI: 10.1038/mt.2015.215. 53

21.   Stark, F. C., Weeratna, R. D., Deschatelets, L., Gurnani, K., Dudani, R., McCluskie, M. J., and Krishnan, L. (2017). An archaeosome-adjuvanted vaccine and checkpoint inhibitor therapy combination significantly enhance protection from murine melanoma, *Vaccine* 5:38. DOI: 10.3390/vaccines5040038. 53

22.   Korsholm. K. S., Hansen, J., Karlsen, K., Filskov, J., Mikkelsen, M., Lindenstrom, T., Schmidt, S. T., Andersen, P., and Christensen, D. (2014). Induction of CD8+ T-cell responses against subunit antigens by the novel cationic liposomal CAF09 adjuvant, *Vaccine* 32:3927–35. DOI: 10.1016/j.vaccine.2014.05.050. 54

23.   Kang, S. W., Lee, S. C., Park, S., H., Kim, J., Kim, H. H., Lee, H. W., Seo, S.-K., Kwon, B. S., Cho, H. R., and Kwon, B. (2017). Anti-CD137 suppresses tumor growth by blocking reverse signaling by CD137 ligand, *Cancer Res.* 77:5989–6000. DOI: 10.1158/0008-5472.CAN-17-0610. 54

24.   Nikpoor, A. R., Tavakkol-Afshari, J., Sadri, K., Jalali, S. A., and Jaafari, M. R. (2017). Improved tumor accumulation and therapeutic efficacy of CTLA-4-blocking antibody using liposome-encapsulated antibody: *In vitro* and *in vivo* studies, *Nanomedicine.* 13:2671–82. DOI: 10.1016/j.nano.2017.08.010. 54

25.   Song, X., Xu, J., Liang, C., Chao, Y., Jin, Q. T., Wang, C., Chen, M., and Liu, Z. (2018). Self-supplied tumor oxygenation through separated liposomal delivery of H2O2 and catalase for enhanced radio-immunotherapy of cancer, *Nano Lett.* 18:6360–8. DOI: 10.1021/acs.nanolett.8b02720. 54

26.   Chandrasekaran, S., Chan, M. F., Li, J. H., and King, M. R. (2016). Super natural killer cells that target metastases in the tumor draining lymph nodes, *Biomaterials* 77:66–76. DOI: 10.1016/j.biomaterials.2015.11.001. 55

27.   Siegler, E. L., Kim, Y. J., Chen, X. H., Siriwon, N., Mac, J., Rohrs, J. A., Bryson, P. D., and Wang, P. (2017). Combination cancer therapy using chimeric antigen receptor-engineered natural killer cells as drug carriers, *Mol. Ther.* 25:2607–19. DOI: 10.1016/j.ymthe.2017.08.010. 55

28.   Pitchaimani. A., Nguyen, T. D. T, and Aryal, S. (2018). Natural killer cell membrane infused biomimetic liposomes for targeted tumor therapy, *Biomaterials* 160:124–37. DOI: 10.1016/j.biomaterials.2018.01.018. 55

29.   Griesmann, H., Drexel, C., Milosevic, N., Sipos, B., Rosendahl, J., Gress, T. M., and Michl, P. (2017). Pharmacological macrophage inhibition decreases metastasis formation in a genetic model of pancreatic cancer, *Gut* 66:1278–85. DOI: 10.1136/gutjnl-2015-310049. 55

30.   Andersen, M. N., Etzerodt, A., Graversen, J. H., Holthof, L. C., Moestrup, S. K., Hokland, M., and Moller, H. J. (2019). STAT3 inhibition specifically in human monocytes and macrophages by CD163-targeted corosolic acid-containing liposomes, *Cancer Immunol. Immunother.* 68:489–502. DOI: 10.1007/s00262-019-02301-3. 55

31.   Kranz, L. M., Diken, M., Haas, H., Kreiter, S., Loquai, C., Reuter, K. C., Meng, M., Fritz, D., Vascotto, F., Hefesha, H., Grunwitz, C., Vormehr, M., Hüsemann, Y., Selmi, A., Kuhn, A. N., Buck, J., Derhovanessian, E., Rae, R., Attig, S., Diekmann, J., Jabulowsky, R. A., Heesch, S., Hassel, J., Langguth, P., Grabbe, S., Huber, C., Özlem, T., and Sahin, U. (2016). Systemic RNA delivery to dendritic cells exploits antiviral defence for cancer immunotherapy, *Nature* 534:396–401. DOI: 10.1038/nature18300. 55

32.   Mitchell, M. J., Jain, R. K., and Langer, R. (2017). Engineering and physical sciences in oncology: challenges and opportunities, *Nat. Rev. Cancer* 17:659–675. DOI: 10.1038/nrc.2017.83. 55

33.   Hua, S., de Matos, M. B. C., Metselaar, J. M., and Storm, G. (2018). Current trends and challenges in the clinical translation of nanoparticulate nanomedicines: pathways for translational development and commercialization, *Front. Pharmacol.* 9:1–14. DOI: 10.3389/fphar.2018.00790. 56

CHAPTER 6

Immunomodulatory Protein Nanoparticles in Cancer Therapy

Vijay Sagar Madamsetty, Anubhab Mukherjee, Sudip Mukherjee

In recent years, chemotherapy and surgery, radiation therapy, and immunotherapy have become the most common treatment options due to failure with motherapy for cancer treatment [1]. Immunotherapy is the safest treatment option among them and has become a fast-developing methodology for cancer treatment [2, 3]. Surgery, chemotherapy, and radiotherapy possess many side effects like damaging normal organs, whereas immunotherapy activates immune cells to detect and eliminate tumor cells without affecting normal organs and tissues [4]. Additionally, by inducing immunological memory cells, the immunotherapeutic approach provides long-term protection against tumor relapse [5]. Recently, immune checkpoint blockade therapy targeting programmed cell death 1 (PD-1), or programmed cell death ligand 1 (PD-L1), and cytotoxic T-lymphocyte antigen 4 (CTLA-4) has elevated extensive consideration in relieving the negative regulation over T cells [6–8]. Recent FDA approval of CAR-T therapy (chimeric antigen receptor T-cell therapy by genetically engineered T cells for antigen-specific tumor therapy) for B-cell and non-Hodgkin lymphoma therapy is also emerging news in the immunotherapy field [9, 10]. However, these immunotherapeutic approaches are limited in treating solid tumors due to abnormal extracellular matrix, extremely immunosuppressive TME, and "off-target" effects of the immune-modulatory agents, which can harm normal tissue cells [11]. These fascinating advancements in cancer immunotherapy are further improved with the support of nanotechnology [12]. Hence, recent nanotechnology advancements can improve cancer immunotherapy's therapeutic efficacy by protecting antigens, especially in nucleic acid, effective delivery to APCs, the origination of potent tumor specific immune response, and the modulating of TME [8]. Among several nanoparticles, protein-based nanoparticle delivery systems directing to moderate immune cells have been established for cancer treatment, and some are in various stages of clinical trials [12, 13]. These advancements define their great potential of immunomodulatory protein nanoparticles as great antitumor agents. In recent studies, protein-based nanoparticles have shown immunostimulating properties [12]. Protein-based nanoparticles can be achieved by the self-assembly of protein structures from various sources other than viruses [14]. These nanoparticles exhibit highly ordered surface shapes and geometries, making them the right delivery platforms for immunotherapy applications [15]. For example, self-assembled protein-based nanoparticles displayed epitopes of the repeat sequence in the circumsporozoite protein elicited a robust immune response against PfCSP [16]. Besides, protein-based nanoparticles that mimic viruses can facilitate DCs cross-presentation and activation [17]. When co-delivered with peptide epitopes, these protein-based nanoparticles presented an improved and prolonged CD8+ T cell activation [18].

E2 constituent of pyruvate dehydrogenase-based self-assembled nanoparticles has become an emergent class of drug delivery vehicles [19]. E2 protein-based nanoparticles are suitable for lymphatic transportation and dendritic cell uptake [20]. DC-activating CpG ODN loaded virus-like DC-targeted nano-vaccine platform was developed for cancer immunotherapy [18]. In another study, CpG ODN in the E2 nano-formulation-activated DCs at a 25-fold lower concentration than free CpG ODN, highlighting the high delivery efficiency of the protein-based nanoparticle approach. Similarly, the diversity of tumor associated antigens has been successfully delivered together with CpG ODN using E2 protein-based nanoparticles for cancer vaccination [21].

Like E2 protein-based nanoformulations, heat-shock proteins- (HSPs) based nanoformulations have also been discovered for cancer immunotherapy [22]. These HSP-based protein nanoparticles display strong receptor-specific interactions with APCs and enable antigen presentation and immune stimulation [23]. Antigenic peptides bound HSP96-based protein nanoparticles showed an excellent antitumor immune response in recurrent GBM (glioblastoma multiforme) and colorectal liver metastases patients [24]. Likewise, melanoma-associated antigen gp100 complexed HSP110-based nanoparticles increased both CD4+ and CD8+ T cell populations in melanoma-bearing mice [25]. Several *in vivo* studies have been conducted on the use of HSP nanoparticles for immunization applications [18]. Similarly, some other ferritin and vault nanoparticle protein-based nanoparticles are also used as natural carriers for antigen delivery [26].

6.1   SYNTHESIS AND CHARACTERIZATION TECHNIQUES

Protein nanoparticles are synthesized by balancing the attractive and repulsive forces in the protein of interest [27]. Increasing protein unfolding and decreasing intramolecular hydrophobic interactions are fundamentaly accepted concepts in forming protein-based nanoparticles [28]. The protein undergoes conformational changes depending on its composition during the formation of the nanoparticles. The formation of protein nanoparticles also mainly depends on its crosslinking ability, concentration, and preparation conditions, including pH, ionic strength, and the solvent used for preparation [29]. In some cases, stabilizing agents such as surfactants are needed to make water-insoluble proteins such as gliadin nanoparticle preparation. Commonly, disulfides and thiols are used as interactive groups to unfold proteins during the preparation process and following chemical or thermal crosslinking leads to the formation of cross-linked protein nanoparticles with entrapped small molecule drugs. Several methods are well established in protein-based nanoparticle preparation; however, coacervation/desolvation and emulsion-based methods are commonly used to prepare protein-based nanoparticles [30, 31].

Coacervation or Desolvation Method

Coacervation or desolvation is the most commonly used method of preparation for protein-based nanoparticles [32]. The coacervation or desolvation preparation method mainly depends on the variance solubility of proteins in solvents, pH, electrolytes, and ionic strength. The coacervation procedure moderates the solubility of the protein, which leads to phase separation. The addition of the desolvating agent's primes to conformational changes in protein structures, subsequent in the protein's coacervation or precipitation. The size of protein-based nanoparticles prepared by this method can control by monitoring the above-discussed variables. The nanoparticle formation stabilized by cross-linking by agents (glutaraldehyde and glyoxal). Mostly, acetone and ethanol are being used as antisolvents for the preparation of protein-based nanoparticles. The selection of solvents depends on the type of protein used for preparing nanoparticles. For example, acetone is the best solvent in preparing smaller size albumin based nanoparticles than in ethanol. The particle size also depends on the ratio between the antisolvent/solvent; increasing these ratio decreases the particle size due to the rapid extraction or diffusion of the solvent into the antisolvent phase, limiting particles' growth.

Similarly, the higher pH conditions produce smaller size nanoparticles, ranging from 100–300 nm. Hence, to get smaller nanoparticles, it is crucial to keep the system's pH away from the pI of protein to promote protein deaggregation. For instance, increasing the BSA protein concentration decreases the formed nanoparticles' size due to their amplified nucleation upon antisolvent addition. Whereas in the preparation of the gelatin-based nanoparticles, gelatin is dissolved in an aqueous solution (pH 7), followed by the altering solvent composition from water to 75% v/v hydroalcoholic solution upon slow ethanol addition under stirring conditions. Cross-linking also plays an important role in stabilizing protein nanoparticles and controlling size. For example, an increase in the crosslinking amount commonly decreases the particle size due to denser particles' formation. Most of the time, the lysine residues in the protein are involved in the crosslinking. Hence, cross-linking stabilizes the protein nanoparticles and decreases enzymatic deprivation and drug release from the formed nanoparticles. However, it is imperative to remove the cross-linkers as much as possible afterward because of their toxicity.

Furthermore, while preparing hydrophobic proteins such as gliadin and legumin-based nanoparticles usually needs surfactants to alleviate the protein nanoparticles. Poloxamer is the most used surfactant to improve legumin's solubility and stabilize the formed nanoparticles. Additionally, increasing the surfactant concentration increases the yield of the nanoparticles without much altering particle size. The loading efficiency depends on drug properties and other factors, such as the drug/polymer ratio. For example, in HSA nanoparticles, a higher drug loading efficiency was reported using this method than the surface adsorption method. In summary, changing the variables dramatically depends on the nature protein used in preparing nanoparticles.

Emulsion or Solvent Extraction

The emulsion or solvent extraction method is also mostly used in preparing protein-based nanoparticles [13]. An aqueous solution of the protein is emulsified in oil using a high-speed homogenizer. Phosphatidylcholine and Span 80 are added as stabilizers in preparing protein based nanoparticles. In the process of making nanoparticles, the oil phase is removed using an organic solvent such as acetone, DCM, and the size of the internal phase determines the ultimate size of nanoparticles. The emulsion-based method is used to prepare a diversity of protein nanoparticles, such as albumin and whey protein nanoparticles. Here, glutaraldehyde is added to the emulsion as a cross-linker to obtain nanoparticles in the range of 100–800 nm in size. The particle size mainly depends on the used protein concentration, emulsification efficiency, and phase volume ratio (w/o), and increasing the protein concentration and phase volume ratio increases the size of the nanoparticles. Sometimes thermal crosslinking is used in place of chemical crosslinking, like to prepare whey protein nanoparticles. In general, the emulsion method-based protein nanoparticles exhibit a larger size than that prepared by the coacervation method. In both cases, removing the oil and organic solvents from the final products is important for the products' safe use.

Complex Coacervation

The complex coacervation method of nanoparticle preparation is ideal for DNA encapsulation for gene therapy applications [33, 34]. Proteins are amphoteric with many charged functional groups, and they can be cationic or anionic by adjusting the pH. These charged proteins can quickly form complexes by electrostatic interactions with DNA or oligonucleotides by coacervation. Salts such as sodium sulfate are sometimes used to induce the polyelectrolyte complex-forming nanoparticles' desolvation, consequently stabilized by crosslinking agents. For example, the salt-induced complex coacervation is using to entrap DNA in gelatin-based protein nanoparticles. At pH 5, gelatin exhibits a positive charge and forms a complex with DNA. In the coacervation process, DNA is physically complexed/entrapped in the protein matrix. Endolysomotropic agents and other drugs can also be co-encapsulated through complex coacervation.

Electrospray

Electrospray is a comparatively new method for the preparation of protein-based nanoparticles [35]. This method is used mostly for the preparation of gliadin and elastin-like peptide-based protein nanoparticles. A high voltage is applied to the protein solution supplied through an emitter during this method, which produces a liquid jet stream through an outlet forming aerosolized liquid droplets. These collected aerosolized droplets contain protein nanoparticles of the colloidal size. Both nucleic acids and drugs can easily be encapsulated into the nanoparticles with high efficacy using this method.

Protein-based nanoparticles synthesized by above methods are characterized by analyzing particle size, PDI, surface area, zeta potential, morphology, and composition. DLS is the main instrument used to measure the developed nanoparticles' size, distribution, and zeta potentials. TEM, SEM, and AFM are also emerging techniques using for the characterization of nanoparticles by looking at their morphological studies. Similarly, some other techniques like FT-IR, XRD, UV-Vis spectrophotometry, HAADF, and ICP are being used in measuring diverse applications of synthesized protein-based nanoparticles.

6.2   RECENT EXAMPLES OF IMMUNOMODULATORY PROTEIN NANOMATERIALS IN CANCER THERAPY

Various drug delivery carriers have been used to improve cancer immune therapy effectiveness with decreasing side effects [36]. Among these carriers, protein-based nanoparticles have been most extensively studied and received the most attention due to highly biodegradable and biocompatible [37]. Due to control in particle size and surface modifications, these nanoparticles have the potential to satisfy the primary requirements of nanocarriers for cancer treatment. Protein-based nanoparticles are relatively safer than others and easy to make, and size can be controlled easily by changing variables [38]. They are flexible to numerous modifications, such as to incorporate functional groups as targeting ligands. PEGylated protein nanoparticles are also being studied as a drug delivery carrier for cancer immunotherapy [39]. Surface modification with PEG has been used to prepare long-circulating gelatin nanoparticles, which exhibit a two-fold increase in plasma level than regular gelatin nanoparticles [39]. We overview the recent developments in the use of engineered protein nanoparticles to enhance cancer immunotherapy.

Vaccine development by natural and synthetic protein-based nanoparticles can potentially be used in biomedicine as vaccines. However, modifications of protein nanoparticles are a viable approach for vaccine development. The encapsulation of target molecules inside protein nanoparticles remains challenging. The nanoparticles' outer surface is chemically or genetically modified to introduce several anchoring points to which antigens can be connected to increase immunogenicity. For example, scientists developed high-density lipoprotein-mimicking protein-based nanodiscs coupled with peptide antigen (Ag) peptides (sHDL-Ag/CpG) for cancer immunotherapy [40]. These sHDL-Ag/CpG protein nanodiscs strikingly increased co-delivery of Ag/CpG to lymphoid organs and sustained Ag presentation on DCs (Figure 6.1). In their study, the authors found that the sHDL-Ag group presented significantly increased FITC signal in draining lymph nodes (dLNs) upon injection of FITC conjugated nanodisc on C57BL/6 mice than free CSSSIINFEK(FITC)L (Figure 6.1a). Likewise, administration of Cy5-labeled Cho-CpG in sHDL nanodisc improved its LN accumulation, compared with a free soluble form (p < 0.01, Figure 6.1b). These results displayed the ability of sHDL nanodisc on delivery of both Ag and CpG to dLNs. Further, the sHDL-Ag/CpG immunized group produced ~21% Ag-specific CD8α + T cells than free Ag peptides and CpG after the third vaccination (Figure 6.1c, d). Similarly, sHDL-Ag/CpG-immunized mice showed no detectable tumors when challenged with 2×10^5 B16OVA cells (Figure 6.1e). In summary, the authors found that long-lived protection against tumor challenge maintains T-cell responses (Figure 6.1), and these protein-based nanodiscs eradicated established B16F10 and MC-38 tumors when combined with anti-CTLA-4 and anti-PD-1 therapy. Their findings epitomize a new powerful protein-based nanoparticle approach for cancer immunotherapy [40]. A similar study was published from the same group in another cancer model [41]. Similarly, a scavenger receptor type B-1 (SCARB1), a high-density lipoprotein (HDL), is highly expressed by myeloid-derived suppressor cells (MDSCs), which potently inhibit T cells population in the tumor microenvironment. Recently, scientists demonstrated synthetic high-density lipoprotein-like nanoparticles that mimic SCARB1 significantly reduced the MDSC activity and enhanced the increased survival with enhancing adaptive immunity [42]. Albumin-based protein nanoparticles are also being using for cancer immunotherapy. In protein-based nanoparticles, albumin-based nanoparticles offer manifold benefits compared to other nanosystems. These possess high biodegradability, biocompatibility, less immunogenicity, and lower cytotoxicity, which are very advantageous features of protein nanocarriers. Paclitaxel-loaded albumin (nab-paclitaxel) is being used to overcome drug resistance and enhance cancer immunotherapy in PDAC. The macropinocytosis uptake of nab-paclitaxel induced macrophage immunostimulatory (M1) cytokine communication and synergized with IFNγ promotes inducible nitric oxide synthase expression in a TLR4-dependent manner [43].

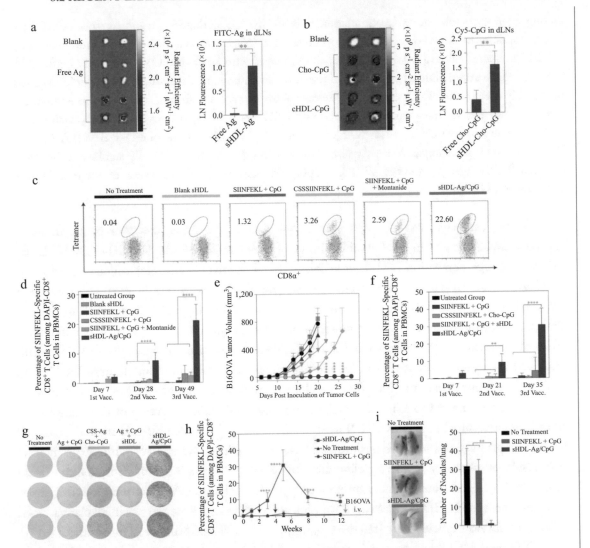

Figure 6.1: Vaccine nanodiscs for LN-targeting of Ag and adjuvants and elicitation of CTL responses. (a,b), C57BL/6 mice were administered subcutaneously at the tail base with 31 nmol FITC-tagged Ag (CSSSIINFEK(FITC)L) (a) or 2.3 nmol Cho-CpG (20% labeled by Cy5) (b) in free soluble or sHDL form, and fluorescence signals in the draining inguinal LNs were quantified with IVIS after 24 h. (c–e), C57BL/6 mice were immunized with the indicated formulations (15.5 nmol Ag peptide and 2.3 nmol CpG) on days 0, 21 and 42. Shown are their representative scatter plots on day 49 (c) and the frequency of SIINFEKL-specific CD8α + T cells in peripheral blood measured 7 days post each immunization by flow-cytometry analysis of tetramer + CD8α + T cells (d). (e), On day 50, pre-vaccinated animals were challenged with subcutaneous flank injection of 2×105 B16OVA cells, and tumor growth was measured over time. (f–h), C57BL/6 mice were immunized in a biweekly interval. Shown are percentage of SIIN-FEKL-specific CD8α + T cells in peripheral blood (f); ELISPOT analysis of IFN-γ spot-forming cells

among splenocytes after ex vivo restimulation with SIINFEKL on day 35 (g); and Ag-specific CD8α + T-cell responses measured over 12 weeks post vaccination (black arrows indicate days of immunizations) (h). (i), Vaccinated mice in h were intravenously challenged with 5×104 B16OVA cells two months after the third vaccination. Shown are pictures of the lungs and numbers of metastatic lung nodules counted on day 20 after the B16OVA challenge. The data show mean ± s.d. from a representative experiment (n=4–5) from 2–3 independent experiments. **p<0.01, ***p<0.001, and ****p<0.0001, analyzed by two-tailed unpaired Student's t-test (a, b), two-way ANOVA (d–f, h), or one-way ANOVA (i) with Bonferroni multiple comparisons post-test. Asterisks in (e) indicate statistically significant dierences between sHDL-Ag/CpG and SIINFEKL + CpG + Montanide. The figure is reproduced after permission from NPG publishers [40].

Tumor-associated macrophages uptake Nab-paclitaxel *in vivo*, and therapeutic doses of nab-paclitaxel alone and in combination with gemcitabine increased the MHCII+CD80+CD86+ M1 macrophage population [43]. Interleukin-2 (IL-2) mediates antitumor cellular immune responses through lymphocyte activation, approved for the IV treatment of renal cell carcinoma (RCC) and melanoma [44]. However, the systemic toxicity and low half-life of rIL-2 to limit its clinical use. Albumin-based delivery approach provides the advantageous pharmacokinetic properties of albumin to a fusion partner (rIL-2), resulted in new protein-based nanoparticles with improved therapeutic potential [45]. In another study, the authors used an albumin hitchhiking approach as a molecular vaccine. They developed amphiphiles (amph-vaccines) containing an antigen cargo connected to a lipophilic albumin-binding tail. Administration of these amph-vaccines in mice resulted in improved LN accumulation, which leads to a 30-fold increase in T-cell population priming and enriched antitumor efficacy with reduced systemic toxicity [46]. Scientists also developed clinically promising albumin/AlbiVax nanocomplexes for efficient albumin-based vaccine delivery and potent cancer immunotherapy. This Albumin/AlbiVax produces ~10 times more antigen-specific CD8+ cytotoxic T lymphocytes with immune memory than regular vaccines. Albumin/AlbiVax explicitly inhibited the progression of established primary or metastatic B16F10, EG7.OVA, and MC38 tumors. Furthermore, in combination with Abraxane and anti-PD-1 therapy improved efficacy immunotherapy and eradicated most MC38 tumors [47]. In another study, scientists developed a Legumin-based nanoparticle to examine the adaptive immune response in rats [48]. Scientists develop numerous protein-based nanoparticles, and more studies are in the process of immunomodulation to enhance cancer immunotherapy [49-52].

6.3   ADVANTAGES AND CHALLENGES

Protein-based nanomaterials have several advantages and applications in delivering agents such as anticancer drugs, growth factors, peptide hormones, genetic materials, RNA, and DNA. These nanoparticles are more stable than other nanomaterials and more comfortable with making in comparison with other types. High possible applications *in vivo* are expected as proteins can be in several types are human-made into nanoparticles using easy methods with low cost and in the green synthesis procedure. Protein-based nanoparticles possess pros and cons, depending on the materials used. Among the countless proteins for drug delivery applications, albumins are most extensively studied. Protein-based nanoparticles can increase protein transfer efficacy by controlling characteristics such as shape, size, and surface charge. Even though protein nanoparticles' application already has some impressive results and has shown a great perspective in the future, comparable data on the protein-based nanoparticles are still limited. Therefore, more protein-based nanoparticles research would be studied. The ideal material or best method must be chosen from material to materials to produce the most effective protein-based nanoparticles.

6.4   REFERENCES

1.   Liu, M. and Guo, F. (2018). Recent updates on cancer immunotherapy, *Precis. Clin. Med.* 1:65-74. DOI: 10.1093/pcmedi/pby011. 61

2.   Stanculeanu, D. L., Daniela, Z., Lazescu, A., Bunghez, R., and Anghel, R. (2016). Development of new immunotherapy treatments in different cancer types, *J. Med. Life* 9:240-8. 61

3.   Waldman, A. D., Fritz, J. M., and Lenardo, M. J. (2020). A guide to cancer immunotherapy: from T cell basic science to clinical practice, *Nat. Rev. Immunol.* 20:651-68. DOI: 10.1038/s41577-020-0306-5. 61

4.   Wirsdörfer, F., de Leve, S., and Jendrossek, V. (2018). Combining radiotherapy and immunotherapy in lung cancer: can we expect limitations due to altered normal tissue toxicity?, *Int. J. Mol. Sci.* 20:24. DOI: 10.3390/ijms20010024. 61

5.   Aldous, A. R. and Dong, J. Z. (2018). Personalized neoantigen vaccines: A new approach to cancer immunotherapy., *Bioorg. Med. Chem.* 26:2842-9. DOI: 10.1016/j.bmc.2017.10.021. 61

6.   Pan, R. Y., Chung, W. H., Chu, M. T., Chen, S. J., Chen, H. C., Zheng, L., and Hung, S. I. (2018). Recent development and clinical application of cancer vaccine: Targeting neoantigens, *J. Immunol. Res.* 4325874. DOI: 10.1155/2018/4325874. 61

7.   Feng, B., Zhou, F., Hou, B., Wang, D., Wang, T., Fu, Y., Ma, Y., Yu, H., and Li, Y. (2018). Binary cooperative prodrug nanoparticles improve immunotherapy by synergistically modulating immune tumor microenvironment, *Adv. Mat.* 30:1803001. DOI: 10.1002/adma.201803001. 61

8.   Feng, X., Xu, W., Li, Z., Song, W., Ding, J., and Chen, X. (2019). Immunomodulatory nanosystems, *Adv. Sci.* (Weinh) 6:1900101. DOI: 10.1002/advs.201900101. 61

9.   Fesnak, A. D., June, C. H., and Levine, B. L. (2016). Engineered T cells: the promise and challenges of cancer immunotherapy, *Nat. Rev. Cancer* 16:566-81. DOI: 10.1038/nrc.2016.97. 61

10.   Seimetz, D., Heller, K., and Richter. J. (2019). Approval of first CAR-Ts: Have we solved all hurdles for ATMPs?, *Cell Med.* 11:2155179018822781. DOI: 10.1177/2155179018822781. 61

11.   Lan, H., Zhang, W., Jin, K., Liu, Y., and Wang, Z. (2020). Modulating barriers of tumor microenvironment through nanocarrier systems for improved cancer immuno-therapy: a review of current status and future perspective, *Drug Del.* 27:1248-62. DOI: 10.1080/10717544.2020.1809559. 61

12.   M,i Y., Hagan Iv, C. T., Vincent, B. G., and Wang, A. Z. (2019). Emerging nano-/microapproaches for cancer immunotherapy, *Adv. Sci.* 6:1801847. DOI: 10.1002/advs.201801847. 61

13.   Lohcharoenkal, W., Wang, L., Chen, Y. C., and Rojanasakul, Y. (2014). Protein nanopar-ticles as drug delivery carriers for cancer therapy, *BioMed. Res. Int.* 180549. DOI: 10.1155/2014/180549. 61, 64

14.   Li, C., Wang, X., Song, H., Deng, S., Li, W., Li, J., and Sun, J. (2020). Current multifunc-tional albumin-based nanoplatforms for cancer multi-mode therapy, *Asian J. Pharm. Sci.* 15:1-12. DOI: 10.1016/j.ajps.2018.12.006. 61

15.   Jain, A., Singh, S. K., Arya, S. K., Kundu, S. C., and Kapoor, S. (2018). Protein nanopar-ticles: Promising platforms for drug delivery applications. *ACS Biomat, Sci. Eng.* 4:3939-61. DOI: 10.1021/acsbiomaterials.8b01098. 61

16.   Kaba, S. A., Karch, C. P., Seth, L., Ferlez, K. M. B., Storme, C. K., Pesavento, D. M., Laughlin, P. Y., Bergmann-Leitner, E. S., Burkhard, P., and Lanar, D. E. (2018). Self-as-sembling protein nanoparticles with built-in flagellin domains increases protective efficacy of a Plasmodium falciparum based vaccine, *Vaccine* 36:906-14. DOI: 10.1016/j.vaccine.2017.12.001. 62

17.   Molino, N. M., Anderson, A. K. L., Nelson, E. L., and Wang, S.-W. (2013). Biomimetic protein nanoparticles facilitate enhanced dendritic cell activation and cross-presentation, *ACS Nano.* 7:9743-52. DOI: 10.1021/nn403085w. 62

18.   Zhuang, J., Holay, M., Park, J. H., Fang, R. H., Zhang, J., and Zhang, L. (2019). Nanoparticle delivery of immunostimulatory agents for cancer immunotherapy, *Theranostics* 9:7826-48. DOI: 10.7150/thno.37216. 62

19.   Patel, M. S., Nemeria, N. S., Furey, W., and Jordan, F. (2014). The pyruvate dehydrogenase complexes: structure-based function and regulation, *J. Biol. Chem.* 289:16615-23. DOI: 10.1074/jbc.R114.563148. 62

20.   Ren, D., Kratz, F., and Wang, S.-W. (2011). Protein nanocapsules containing doxorubicin as a pH-responsive delivery system, *Small* 7:1051-60. DOI: 10.1002/smll.201002242. 62

21.   Lim, S., Park, J., Shim, M. K., Um, W., Yoon, H. Y., Ryu, J. H., Lim, D. K., and Kim, K. (2019). Recent advances and challenges of repurposing nanoparticle-based drug delivery systems to enhance cancer immunotherapy, *Theranostics* 9:7906-23. DOI: 10.7150/thno.38425. 62

22.   Lin, F.-C., Hsu, C.-H., and Lin, Y.-Y. (2018). Nano-therapeutic cancer immunotherapy using hyperthermia-induced heat shock proteins: insights from mathematical modeling, *Int. J. Nanomed.* 13:3529-39. DOI: 10.2147/IJN.S166000. 62

23.   Shevtsov, M. and Multhoff, G. (2016). Heat shock protein-peptide and HSP-based immunotherapies for the treatment of cancer, *Front Immunol.* 7:171-. DOI: 10.3389/fimmu.2016.00171. 62

24.   Crane, C. A., Han, S. J., Ahn, B., Oehlke, J., Kivett, V., Fedoroff, A., Butowski, N., Chang, S. M., Clarke, J., Berger, M. S., McDermott, M. W., Prados, M. D., and Parsa, A. T. (2013). Individual patient-specific immunity against high-grade glioma after vaccination with autologous tumor derived peptides bound to the 96 KD chaperone protein, *Clin. Cancer Res.* 19:205. DOI: 10.1158/1078-0432.CCR-11-3358. 62

25.   Wang, X. Y., Chen, X., Manjili, M. H., Repasky, E., Henderson, R., and Subjeck, J. R. (2003). Targeted immunotherapy using reconstituted chaperone complexes of heat shock protein 110 and melanoma-associated antigen gp100, *Cancer Res.* 63:2553-60. 62

26.   Diaz, D., Care, A., and Sunna, A. (2018). Bioengineering strategies for protein-based nanoparticles, *Genes* (Basel) 9:370. DOI: 10.3390/genes9070370. 62

27.   Lohcharoenkal, W., Wang, L., Chen, Y. C., and Rojanasakul, Y. (2014). Protein nanoparticles as drug delivery carriers for cancer therapy, *BioMed. Res. Int.* 180549-. DOI: 10.1155/2014/180549. 62

28.   Zink, J., Wyrobnik, T., Prinz, T., and Schmid, M. (2016). Physical, chemical and biochemical modifications of protein-based films and coatings: An extensive review, *Int. J. Mol. Sci.* 17:1376. DOI: 10.3390/ijms17091376. 62

29.   Tarhini, M., Benlyamani, I., Hamdani, S., Agusti, G., Fessi, H., Greige-Gerges, H., Bentaher, A., and Elaissari, A. (2018). Protein-based nanoparticle preparation via nano-precipitation method, *Materials* (Basel) 11:394. DOI: 10.3390/ma11030394. 62

30.   Hong, S., Choi, D. W., Kim, H. N., Park, C. G., Lee, W., and Park, H. H. (2020). Protein-based nanoparticles as drug delivery systems, *Pharmaceutics* 12:604. DOI: 10.3390/pharmaceutics12070604. 63

31.   Lohcharoenkal, W., Wang, L., Chen, Y. C., and Rojanasakul, Y. (2014). Protein nanoparticles as drug delivery carriers for cancer therapy, *Biomed. Res. Int.* 180549. DOI: 10.1155/2014/180549. 63

32.   Verma, D., Gulati, N., Kaul, S., Mukherjee, S., and Nagaich, U. (2018). Protein based nanostructures for drug delivery, *J. Pharm.* (Cairo) 9285854-. DOI: 10.1155/2018/9285854. 63

33.   Müller, W. E. G., Tolba, E., Wang, S., Neufurth, M., Lieberwirth, I., Ackermann, M., Schröder, H. C., and Wang, X. (2020). Nanoparticle-directed and ionically forced polyphosphate coacervation: a versatile and reversible core–shell system for drug delivery, *Sci. Rep.* 10:17147. DOI: 10.1038/s41598-020-73100-5. 64

34.   Fuchs, S. and Coester, C. (2010). Protein-based nanoparticles as a drug delivery system: chances, risks, perspectives, *J. Drug Del. Sci. Tech.* 20:331-42. DOI: 10.1016/S1773-2247(10)50056-X. 64

35.   Sridhar, R. and Ramakrishna, S. (2013). Electrosprayed nanoparticles for drug delivery and pharmaceutical applications, *Biomatter* 3:e24281. DOI: 10.4161/biom.24281. 64

36.   Lim, S., Park, J., Shim, M. K., Um, W., Yoon, H. Y., Ryu, J. H., Lim, D.-K., and Kim, K. (2019). Recent advances and challenges of repurposing nanoparticle-based drug delivery systems to enhance cancer immunotherapy, *Theranostics* 9:7906-23. DOI: 10.7150/thno.38425. 65

37.   DeFrates, K., Markiewicz, T., Gallo, P., Rack, A., Weyhmiller, A., Jarmusik, B., and Hu, X. (2018). Protein polymer-based nanoparticles: Fabrication and medical applications, *Int. J. Mol. Sci.* 19:1717. DOI: 10.3390/ijms19061717. 65

38.   Yu, M., Wu, J., Shi, J., and Farokhzad, O. C. (2016). Nanotechnology for protein delivery: Overview and perspectives, *J. Control Rel.* 240:24-37. DOI: 10.1016/j.jconrel.2015.10.012. 65

39.   Mishra, P., Nayak, B., and Dey, R. K. (2016). PEGylation in anti-cancer therapy: An overview, *Asian J. Pharm. Sci.* 11:337-48. DOI: 10.1016/j.ajps.2015.08.011. 65

40. Kuai, R., Ochyl, L. J., Bahjat, K. S., Schwendeman, A., and Moon, J. J. (2017). Designer vaccine nanodiscs for personalized cancer immunotherapy, *Nat. Mat.* 16:489-96. DOI: 10.1038/nmat4822. 65, 68

41. Kuai, R., Sun, X., Yuan, W., Xu, Y., Schwendeman, A., and Moon, J. J. (2018). Subcutaneous nanodisc vaccination with neoantigens for combination cancer immunotherapy, *Bioconjug. Chem.* 29:771-5. DOI: 10.1021/acs.bioconjchem.7b00761. 66

42. Plebanek, M. P., Bhaumik, D., Bryce, P. J., and Thaxton, C. S. (2018). Scavenger receptor type b1 and lipoprotein nanoparticle inhibit myeloid-derived suppressor cells, *Mole. Cancer Therapeut.* 17:686. DOI: 10.1158/1535-7163.MCT-17-0981. 66

43. Cullis, J., Siolas, D., Avanzi, A., Barui, S., Maitra, A., and Bar-Sagi, D. (2017). Macropinocytosis of nab-paclitaxel drives macrophage activation in pancreatic cancer, *Cancer Immunol. Res.* 5:182-90. DOI: 10.1158/2326-6066.CIR-16-0125. 66, 68

44. Rosenberg, S. A. (2001). Progress in human tumor immunology and immunotherapy, *Nature* 411:380-4. DOI: 10.1038/35077246. 68

45. Van de Sande, L., Cosyns, S., Willaert, W., and Ceelen, W. (2020). Albumin-based cancer therapeutics for intraperitoneal drug delivery: A review, *Drug Deliv.* 27:40-53. DOI: 10.1080/10717544.2019.1704945. 68

46. Liu, H., Moynihan, K. D., Zheng, Y., Szeto, G. L., Li, A. V., Huang, B., Van Egeren, D. S., Park, C., and Irvine, D. J. (2014). Structure-based programming of lymph-node targeting in molecular vaccines, *Nature* 507:519-22. DOI: 10.1038/nature12978. 68

47. Zhu, G., Lynn, G. M., Jacobson, O., Chen, K., Liu, Y., Zhang, H., Ma, Y., Zhang, F., Tian, R., Ni, Q., Cheng, S., Wang, Z., Lu, N., Yung, B. C., Wang, Z., Lang, L., Fu, X., Jin, A., Weiss, I. D., Vishwasrao, H., Niu, G., Shroff, H., Klinman, D. M., Seder, R. A., and Chen, X. (2017). Albumin/vaccine nanocomplexes that assemble *in vivo* for combination cancer immunotherapy, *Nat. Commun.* 8:1954. DOI: 10.1038/s41467-017-02191-y. 68

48. Mirshahi, T., Irache, J. M., Nicolas, C., Mirshahi, M., Faure, J. P., Gueguen, J., Hecquet, C., and Orecchioni, A. M. (2002). Adaptive Immune Responses of Legumin Nanoparticles, *J. Drug Target..* 10 :625-31. DOI: 10.1080/1061186021000066237. 68

49. Hotaling, N. A., Tang, L., Irvine, D. J., and Babensee, J. E. (2015). Biomaterial strategies for immunomodulation, *Ann. Rev. Biomed. Eng.* 17:317-49. DOI: 10.1146/annurev-bioeng-071813-104814. 68

50. Irvine, D. J. and Dane, E. L. (2020). Enhancing cancer immunotherapy with nanomedicine, *Nat. Rev. Immunol.* 20:321-34. DOI: 10.1038/s41577-019-0269-6. 68

51.   Park, W., Heo, Y.-J., and Han, D. K. (2018). New opportunities for nanoparticles in cancer immunotherapy, *Biomat. Res.* 22:24. DOI: 10.1186/s40824-018-0133-y. 68

52.   Locy, H., de Mey, S., de Mey, W., De Ridder, M., Thielemans, K., and Maenhout, S. K. (2018). Immunomodulation of the tumor microenvironment: Turn foe into friend, *Front. Immunol.* 9. DOI: 10.3389/fimmu.2018.02909. 68

CHAPTER 7

Immunomodulatory Viral Nanoparticles in Cancer Therapy

Vijay Sagar Madamsetty, Anubhab Mukherjee, Sudip Mukherjee

Cancer is a deadly disease of uncontrolled cell growth and spread of abnormal cells [1]. Current conventional cancer therapies, including surgery, chemotherapy, and radiation, have significantly improved over a few years [2]. However, the efficient treatment options for cancer remain a forbidden challenge. Even though the human immune system can fight against several diseases, including cancer, cancer cells divide and mutate rapidly, so the ability of the natural immune system to target cancerous cells is greatly diminished [3]. However, so far, the clinical effect of most cancer immunotherapies are not satisfactory. Hence, there is a great need to develop efficient immune therapies for the betterment of cancer treatment.

It has been widely recognized that nanotechnology can enhance the efficacy of cancer immunotherapy and recently emerged with promising results in the healthcare department [4–6]. Using the benefits of nanotechnology and understanding the fundamental mechanisms behind the human immune response against cancer cells can develop efficient immunomodulatory nanodrugs for cancer immunotherapies [7, 8]. The cancer immune surveillance process is a knowing mechanism that produces the immune cells to identify and attach to cancer-specific target cell antigens [9, 10]. Antigen-attached antibody activates the immune system machinery to start target cell devastation. A variety of nano-based immunotherapies have been developed that have excellent modifiability and immunogenicity [11]. These nanoparticles can carry tumor therapeutic drugs and immunomodulatory agents to achieve combined therapy to improve antitumor immunity's efficiency with reducing adverse side effects.

Among several kinds of nanoparticles for cancer immunotherapy treatment, viral-like nanoparticles (VLNP) recently gained more attention because of VLNP's boost immunotherapy against cancer [12, 13]. More importantly, these virus nanoparticles can carry or express multivalent antigens to realize multi-antigens specific adaptive immunity [12]. Natural viruses can attack a host organism's immune system to warrant its own survival and replica [14]. Virus-like particles based nanoparticles (VLNPs) are safe and economic tools for treating several diseases, including cancer. Hence, these structures are devoid of viral genetic material and are not capable of replicating or causing any infection. VLNPs are capable of carrying the genetic information required for immunotherapy [12].

Several viral vectors have been tested for cancer therapy for boosting the immune system, such as the animal virus adenovirus and the plant virus Cowpea mosaic virus (CPMV) [15]. Plant viruses are the most established and extensively used for cancer vaccine development and are valuable alternative nanoparticles for drug delivery [12, 15, 16]. The successful results of plant viruses in cancer treatment have been established using Hibiscus chlorotic ringspot virus, Tomato bushy stunt virus, and Red clover necrotic mosaic virus [17]. Plant virus-based nanoparticles offer the advantage of uniformity concerning size and shape and self assemble into highly repeating nanostructures [18]. Plant VLNPs also exhibit structurally well-defined chemical attachment sites, protect from high pH, temperature, proteases, nucleases found in the intracellular environment, and a payload capacity suitable for transporting anticancer drugs to disease sites using assemblies of the plant-virus capsid protein [19-21] .

Viruses are non-cellular, tiny organisms compared to other living organisms and assemble from pools of their structural components [12, 22]. They make biotrophic parasites on their hosts and use host proteins for their own survival purposes [22]. Virus particles are composed of two constituents, an infective core of nucleoproteins, and a surrounding protective protein shell. Hence, synthesized viral nanoparticles are highly symmetrical, dynamic, polyvalent, self-assembling, and monodisperse. Naturally occurring virus-based bionanomaterials are replication-deficient, biodegradable, and biocompatible in mammals, including humans [23]. The size of many viruses is less than 500 nm, can smoothly perform chemical alterations, and high tolerance to pH, temperature, and organic solvent-water mixtures [12, 22, 24, 25]. That is why virus-based nanoparticles are considered as potential platforms for the development of nanomaterials in the fields of nanomedicine. There are several types of viruses used in developing nanoparticles such as Brome mosaic virus (BMV), Cowpea chlorotic mottle virus (CCMV), Carnation mottle virus (CarMV), Cowpea mosaic virus (CPMV), Maize rayado fino virus (MRFV), Hibiscus chlorotic ringspot virus (HCRSV), Red clover necrotic mottle virus (RCNMV), Turnip yellow mosaic virus (TYMV), and Sesbania mosaic virus (SeMV) [13, 26, 27]. Viral nanoparticles can be synthesized mostly by self-assembly methods and perform many modifications such as PEGylation, antibody conjugation, etc. [25, 28–30]. Plant leaves were isolated and homogenized, ultracentrifuge followed by gradient ultracentrifugation and size-exclusion fast protein liquid chromatography (FPLC) for producing plant virus-based nanoparticles. CPMV based nanoparticles are produced using an *in vitro* mixed re-assembly method [31]. In this approach, two capsid populations are functionalized with different display ligands are mixed in a specific ratio and allowed to re-assemble together to form a mosaic capsid [32]. Much investigation has been published on the production and use of synthetic, mosaic, or hybrid virus-particles that self-assembles viral capsid proteins and chemically reactive polymers [33]. VLNPs have also been made from mixing virus subunits and synthetic polymers such as the dendrimer polyamidoamine. Like other nanoparticles, the synthesized virus-based nanoparticles are also characterized by investigating size, PDI, zeta potential, surface area, and morphology by using DLS, SEM), TEM, etc. [25, 34–37].

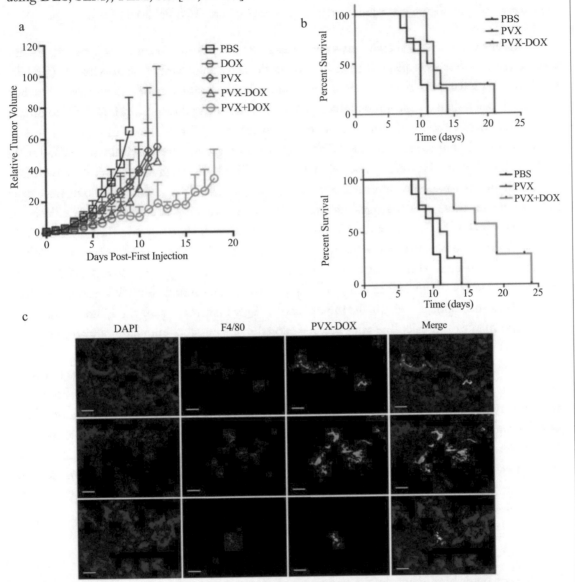

Figure 7.1: Chemo-immunotherapy treatment of B16F10 tumors. Groups (n = 6) were treated with PBS, PVX, DOX, PVX–DOX, or PVX+DOX. PVX was administered at a dose of 5 mg kg–1, DOX was administered at a dose of 0.065 mg kg–1. Treatment started ~8 days post induction when tumors measured 1000 mm^3. (a) Tumor growth curves are shown as relative tumor volume. Statistical significance was detected comparing PVX vs. PVX+DOX. (b) Survival rates of treated mice. (c) Immunofluorescence imaging of three representative PVX –DOX tumor sections after weekly dosing of PVX–DOX (animals

received two doses of PVX and were collected when tumors reached >1000 mm^3). Tumors treated with PVX –DOX (rows 1–3) were sectioned and stained with DAPI (blue), F4/80 (red), and PVX (green). Scale bar = 100 μm. Figure was reproduced after permission from ACS Publishers [38].

Recent studies have shown that some viruses like the flexuous plant virus are using in-situ vaccine therapy. For example, scientists developed flexuous plant virus (potato virus X (PVX))-based viral nanoparticles used as an immunotherapeutic for in situ vaccine monotherapy. Further, they developed a dual chemo-immunotherapeutics by integrating doxorubicin into PVX to obtain a dual-functional virus-based nanoparticle to treat melanoma [38]. As shown in Figures 7.1a and b, the authors did not see any statistical difference in tumor growth and survival with PVX-DOX, PVX, or DOX alone. However, the combination of PVX+DOX treatment significantly inhibited the tumor growth rate and improved survival compared with PVX and DOX alone treatment [38]. Therefore, these results suggesting that PVX immunotherapy is enhancing the efficacy of chemotherapy (DOX). However, the PVX–DOX formulation did not improve the treatment. This is explained by the fact that some therapies synergize best when they act on their own. PVX is most likely connected with immune cells to stimulate an antitumor effect, whereas DOX targets replicating cancer cells to induce cell death [38]. Further, they found that PVX was colocalized with F4/80+ macrophages within the tumor tissue [38], which may cause the killing of immune cells rather than cancer cells. The same group explored many possible immunomodulatory roles of virus-based nanoparticles for cancer immunotherapy [39–41].

Overall, virus-based nanoparticles showed potential as immunomodulatory agents in cancer therapy. These are generally safe and offer highly multipurpose platforms that can be manipulated in specificity and moieties' presentation for nanotechnological and therapeutic applications. However, enormous research was done using viral nanomaterials on in tissue culture *in vitro*, and only a few systems have been evaluated *in vivo*. Hence, developments of novel bio nanoparticles using new viruses are being expedited to capitalize on the abilities of new viral multivalent structures for immunomodulatory agents. Viral-based nanoparticle research is still relatively less explored, and more *in vivo* evaluations are needed for a better understanding.

7.1   CONCLUSION AND FUTURE OUTLOOK

This present book highlights various immunomodulatory nanoparticles that are currently being used for cancer therapeutic applications. It is important to note that some of these technology are in the initial developmental stage and a detailed study is required regarding their toxicity, pharmacokinetics and pharamacodynamics before their clinical translation. The future in this field is really bright and several nanomedicine-based solutions will be clinically approved in the near future.

7.2   REFERENCES

1.   Hanahan, D. and Weinberg, R. A. (2011). Hallmarks of cancer: The next generation, *Cell* 144:646–74. DOI: 10.1016/j.cell.2011.02.013. 75

2.   Arruebo, M., Vilaboa, N., Sáez-Gutierrez, B., Lambea, J., Tres, A., Valladares, M., and González-Fernández, A. (2011). Assessment of the evolution of cancer treatment therapies, *Cancers* (Basel) 3:3279–330. DOI: 10.3390/cancers3033279. 75

3.   Gonzalez, H., Hagerling, C., and Werb, Z. (2018). Roles of the immune system in cancer: from tumor initiation to metastatic progression, *Genes Dev.* 32:1267–84. DOI: 10.1101/gad.314617.118. 75

4.   Cheng, K., Kang, Q., and Zhao, X. (2020). Biogenic nanoparticles as immunomodulator for tumor treatment, *WIREs Nanomed. Nanobiotech.* 12:e1646. DOI: 10.1002/wnan.1646. 75

5.   Huang, Y. and Zeng, J. (2020). Recent development and applications of nanomaterials for cancer immunotherapy, *Nanotech. Rev.* 9:367–84. DOI: 10.1515/ntrev-2020-0027. 75

6.   Zhuang, J., Holay, M., Park, J. H., Fang, R. H., Zhang, J., and Zhang, L. (2019). Nanoparticle delivery of immunostimulatory agents for cancer immunotherapy, *Theranostics* 9:7826–48. DOI: 10.7150/thno.37216. 75

7.   Feng, X., Xu, W., Li, Z., Song, W., Ding, J., and Chen, X. (2019). Immunomodulatory nanosystems, *Adv. Sci.* (Weinh) 6:1900101. DOI: 10.1002/advs.201900101. 75

8.   Kumar Selvaraj, V. and Kumar Gudipudi, D. (2020). Fundamentals to clinical application of nanoparticles in cancer immunotherapy and radiotherapy, *ecancermedicalscience* 14. DOI: 10.3332/ecancer.2020.1095. 75

9.   Pardoll, D. (2015). Cancer and the immune system: basic concepts and targets for intervention, *Semin. Oncol.* 42:523–38. DOI: 10.1053/j.seminoncol.2015.05.003. 75

10.   Makkouk, A. and Weiner, G. J. (2015). Cancer immunotherapy and breaking immune tolerance: new approaches to an old challenge, *Cancer Res.* 75:5–10. DOI: 10.1158/0008-5472.CAN-14-2538. 75

11.   Mi, Y., Hagan IV, C. T., Vincent, B. G., and Wang, A. Z. (2019). Emerging nano-/microapproaches for cancer immunotherapy, *Adv. Sci.* (Weinh) 6:1801847. DOI: 10.1002/advs.201801847. 75

12.   Chung, Y. H., Cai, H., and Steinmetz, N. F. (2020). Viral nanoparticles for drug delivery, imaging, immunotherapy, and theranostic applications, *Adv. Drug Deliv. Rev.* S0169-409X(20)30070-3. 75, 76

13.   Beatty, P. H. and Lewis, J. D. (2019). Cowpea mosaic virus nanoparticles for cancer imaging and therapy, *Adv. Drug Deliv. Rev.* 145:130–44. DOI: 10.1016/j.addr.2019.04.005. 75, 76

14.   DeFilippis, V. R. and Villarreal, L. P. (2000). An introduction to the evolutionary ecology of viruses, *Viral Ecol.* 125–208. DOI: 10.1016/B978-012362675-2/50005-7. 75

15.   Shoeb, E. and Hefferon, K. (2019). Future of cancer immunotherapy using plant virus-based nanoparticles, *Future Sci. OA* 5 FSO401–FSO. DOI: 10.2144/fsoa-2019-0001. 76

16.   Madamsetty, V. S., Mukherjee, A., and Mukherjee, S. (2019). Recent trends of the bio-inspired nanoparticles in cancer theranostics, *Frontiers Pharmacol.* 10. DOI: 10.3389/fphar.2019.01264. 76

17.   Franzen, S. and Lommel, S. A. (2009). Targeting cancer with 'smart bombs': equipping plant virus nanoparticles for a 'seek and destroy' mission, *Nanomedicine* 4:575–88. DOI: 10.2217/nnm.09.23. 76

18.   Zhang, Y., Dong, Y., Zhou, J., Li, X., and Wang, F. (2018). Application of plant viruses as a biotemplate for nanomaterial fabrication, *Molecules* 23:2311. DOI: 10.3390/molecules23092311. 76

19.   Matsuura, K. (2018). Synthetic approaches to construct viral capsid-like spherical nanomaterials, *Chem. Comms.* 54:8944–59. DOI: 10.1039/C8CC03844A. 76

20.   Wu, J., Wu, H., Nakagawa, S., and Gao, J. (2020). Virus-derived materials: bury the hatchet with old foes, *Biomater. Sci.* 8:1058–72. DOI: 10.1039/C9BM01383K. 76

21.   Wang, C., Beiss, V., and Steinmetz, N. F. (2019). Cowpea mosaic virus nanoparticles and empty virus-like particles show distinct but overlapping immunostimulatory properties, *J. Virol.* 93:e00129–19. DOI: 10.1128/JVI.00129-19. 76

22.   Koonin, E. V. and Dolja, V. V. (2014). Virus world as an evolutionary network of viruses and capsidless selfish elements., *Microbiol. Mol. Biol. Rev.* 78:278–303. DOI: 10.1128/MMBR.00049-13. 76

23.   Steinmetz, N. F. (2010). Viral nanoparticles as platforms for next-generation therapeutics and imaging devices., *Nanomedicine* 6:634–41. DOI: 10.1016/j.nano.2010.04.005. 76

24.   Wen, A. M., Lee, K. L., Yildiz, I., Bruckman, M. A., Shukla, S., and Steinmetz, N. F. (2012). Viral nanoparticles for *in vivo* tumor imaging, *J. Vis. Exp.* e4352. DOI: 10.3791/4352. 76

25. Cho, C.-F., Shukla, S., Simpson, E. J., Steinmetz, N. F., Luyt, L. G., and Lewis, J. D. (2014). Molecular targeted viral nanoparticles as tools for imaging cancer, *Methods Mol. Biol.* 1108:211–30. DOI: 10.1007/978-1-62703-751-8_16. 76, 77

26. Narayanan, K. B. and Han, S. S. (2017). Icosahedral plant viral nanoparticles - bioinspired synthesis of nanomaterials/nanostructures, *Adv. Colloid. Interf. Sci.* 248:1–19. DOI: 10.1016/j.cis.2017.08.005. 76

27. Zhang, W., Xu, C., Yin, G. Q., Zhang, X. E., Wang, Q., and Li, F. (2017). Encapsulation of inorganic nanomaterials inside virus-based nanoparticles for bioimaging, *Nanotheranostics* 1(4):358–368. DOI: 10.7150/ntno.21384. 76

28. (2018). Virus-Derived Nanoparticles for Advanced Technologies. 76

29. Pokorski, J. K. and Steinmetz, N. F. (2011). The art of engineering viral nanoparticles, *Mol. Pharm.* 8:29–43. DOI: 10.1021/mp100225y. 76

30. Lee, K. L., Uhde-Holzem, K., Fischer, R., Commandeur, U., and Steinmetz, N. F. (2014). Genetic engineering and chemical conjugation of potato virus X, *Meth. Mol. Biol.* 1108:3–21. DOI: 10.1007/978-1-62703-751-8_1. 76

31. Leong, H. S., Steinmetz, N. F., Ablack, A., Destito, G., Zijlstra, A., Stuhlmann, H., Manchester, M., and Lewis, J. D. (2010). Intravital imaging of embryonic and tumor neovasculature using viral nanoparticles, *Nat. Protoc.* 5:1406–17. DOI: 10.1038/nprot.2010.103. 76

32. Koudelka, K. J., Pitek, A. S., Manchester, M., and Steinmetz, N. F. (2015). Virus-based nanoparticles as versatile nanomachines, *Annu. Rev. Virol.* 2:379–401. DOI: 10.1146/annurev-virology-100114-055141. 76

33. Wen, A. M. and Steinmetz, N. F. (2016). Design of virus-based nanomaterials for medicine, biotechnology, and energy, *Chem. Soc. Rev.* 45:4074–126. DOI: 10.1039/C5CS00287G. 76

34. Destito, G., Schneemann, A., and Manchester, M. (2009). *Viruses and Nanotechnology*, Eds. M. Manchester and N. F. Steinmetz (Berlin, Heidelberg: Springer Berlin Heidelberg) pp. 95–122. DOI: 10.1007/978-3-540-69379-6_5. 77

35. Oh, J.-W. and Han, D.-W. (2020). *Virus-Based Nanomaterials and Nanostructures*, Multidisciplinary Digital Publishing Institute). DOI: 10.3390/nano10030567. 77

36. Han, D.-W. and Oh, J.-W. (2020). *Virus-Based Nanomaterials and Nanostructures* (MDPI). 77

37.   Ren, Y., Wong, S. M., and Lim, L.-Y. (2007). Folic acid-conjugated protein cages of a plant virus: A novel delivery platform for doxorubicin, *Bioconjugate Chem.* 18:836–43. DOI: 10.1021/bc060361p. 77

38.   Lee, K. L., Murray, A. A., Le, D. H. T., Sheen, M. R. Shukla, S., Commandeur, U., Fiering, S., and Steinmetz, N. F. (2017). Combination of plant virus nanoparticle-based in situ vaccination with chemotherapy potentiates antitumor response, *Nano Lett.* 17:4019–28. DOI: 10.1021/acs.nanolett.7b00107. 78

39.   Murray, A. A., Wang, C., Fiering, S., and Steinmetz, N. F. (2018). In situ vaccination with cowpea vs tobacco mosaic virus against melanoma, *Mol. Pharm.* 15:3700–16. DOI: 10.1021/acs.molpharmaceut.8b00316. 78

40.   Le, D. H. T., Commandeur, U., and Steinmetz, N. F. (2019). Presentation and delivery of tumor necrosis factor-related apoptosis-inducing ligand via elongated plant viral nanoparticle enhances antitumor efficacy, *ACS Nano.* 13:2501–10. DOI: 10.1021/acsnano.8b09462. 78

41.   Le, D. H., Lee, K. L., Shukla, S., Commandeur, U., and Steinmetz, N. F. (2017). Potato virus X, a filamentous plant viral nanoparticle for doxorubicin delivery in cancer therapy, *Nanoscale* 9:2348–57. DOI: 10.1039/C6NR09099K. 78

Authors' Biographies

Dr. Anubhab Mukherjee completed his Ph.D. from the CSIR-Indian Institute of Chemical Technology, Hyderabad, India, having been skilled with RNAi, liposomal drug delivery in cancer, preclinical cell, and animal studies. He pursued post-doctoral research at the College of Pharmacy, Health Science Center, Texas A&M University, and was involved in another postdoctoral research at Department of Translational Neurosciences and Neurotherapeutics, John Wayne Cancer Institute, Santa Monica, California. In 2015, he worked at the Harvard-MIT Health Sciences and Technology as a visiting scientist. He has substantial experience in nanotechnology-based formulation development and successfully worked for various Indian organizations to develop pharmaceuticals and nutraceutical formulations. He is an inventor of many U.S. patents and an author of many peer-reviewed articles published in various international journals of repute. Dr. Mukherjee is currently serving as Principal Scientist, R&D at Esperer Onco Nutrition (EON) Pvt. Ltd. and heads the Hyderabad R&D center of the organization.

Dr. Vijay Sagar Madamsetty received his Ph.D. in chemical science from the Academy of Scientific and Innovative Research (AcSIR), India. He has had a passion for developing cutting-edge research on nanomedicine for about ten years, starting from the Council of Scientific and Industrial Research-Indian Institute of Chemical Technology (CSIR-IICT) and continuing to the present as a senior postdoctoral fellow at the Mayo Clinic Florida in the Department of Biochemistry and Molecular Biology. He is currently developing multiple drugs/genes encapsulating targeted nanoformulations and their preclinical evaluation in several disease animal models. Dr. Madamsetty's long-term research interests involve understanding key signaling pathways and how alterations in gene expression contribute to human diseases, leading to developing effective targeted nanomedicine. He is also an active reviewer and editorial board member of several international journals.

Dr. Sudip Mukherjee completed his Ph.D. from the CSIR-Indian Institute of Chemical Technology, Hyderabad, India. Dr. Mukherjee is currently working as a Postdoctoral Research Associate at Rice University. His research is involved in the development of advanced nanomaterials for drug/gene delivery in cancer theranostics, immunomodulatory applications, and angiogenesis. He has published a total of ~50 research articles/patents. He serves as International Advisory Board Member for "Materials Research Express," IOP Sciences. He is a member (MRSC) of RSC, UK. He serves as reviewer for several international journals such as *ChemComm*, *Journal of Materials Chemistry A*, *Journal of Materials Chemistry B*, *Journal of Biomedical Nanotechnology*, *RSC Advances*, and *IOP Nanotechnology*, *Biofabrication*.

Printed in the United States
by Baker & Taylor Publisher Services